HOW TO REBUILD BIG-BLOCK MOPAR ENGINES

Covers all years and models of B/RB Series Chrysler, Dodge and Plymouth engines

DON TAYLOR

HPBooks

HPBooks
are published by
The Berkley Publishing Group
A division of Penguin Putnam Inc.
375 Hudson Street
New York, New York 10014

First edition: October 1994

19 18 17 16 15 14 13 12 11

The Penguin Putnam Inc. World Wide Web site address is http://www.penguinputnam.com

Library of Congress Cataloging-in-Publication Data

Taylor, Don (Donald D.)
How to rebuild big-block Mopar engines / don taylor,
p. cm.
Includes index.
ISBN 1-55788-190-1 (pbk.): $17.00
1. Automobiles—Motors—Modification. 2. Chrysler automobile.
I. Title.
TL210.T378 1994
629.25′04—do20 94-17342

Book Design & Production by Bird Studios.
Cover photo courtesy *Mopar Muscle* magazine. Interior photos by Don Taylor.

NOTICE: The information in this book is true and complete to the best of our knowledge. All recommendations on parts and procedures are made without any guarantees on the part of the author or The Berkley Publishing Group. Tampering with, altering, modifying or removing any emissions control device is a violation of federal law. Author and publishers disclaim all liability incurred in connection with the use of this information.

CONTENTS

ACKNOWLEDGMENTS iv

INTRODUCTION vi

CHAPTER 1
INSPECTION 1

CHAPTER 2
ENGINE REMOVAL 13

CHAPTER 3
DISASSEMBLY 29

CHAPTER 4
PARTS IDENTIFICATION & INTERCHANGE 47

CHAPTER 5
RECONDITIONING THE SHORT BLOCK 65

CHAPTER 6
RECONDITIONING THE HEADS 91

CHAPTER 7
ENGINE ASSEMBLY 115

CHAPTER 8
INSTALLATION 145

INDEX 164

ACKNOWLEDGMENTS

This is the best part of the book for the author. Here, he gets to pretend he's just received the Pulitzer Prize and now gets to make a speech to all the world thanking those who helped make the book such a success, and indeed there have been many!

No one person has all the information. No one person is the complete authority—"the walking encyclopedia." So, to make this book what it is, many, many friends gave liberally of their time and knowledge.

Doing everything they could to help me find information and contact sources were the great team at *Chrysler Power* magazine. This dedicated group is led by Roland Osborne, Publisher and all 'round good guy. Making sure he does everything right though, are Carrie Hynes, Operations Manager; Pete Figueroa, Senior Engineer; Rick Thornton, Sales Support; and Jason Muckala, Technical Support. Thanks, friends. I'd still be looking for that chrome-plated exhaust bearing if you hadn't set me straight!

If you want the best in machine shop work you won't have to look further than Gene Ohly's team at Evans Speed Equipment in South El Monte, California. This company's been in business over 40 years so I had no qualms about their ability to provide the best machine work available. Gene is ably assisted by Dick New, the numero uno machinist, and Jeff Elgin, a fine young man coming up in the business. Janet McKeehan answers the phone and helps keep those miles of paper work in order. Together, this team made it possible to get all of the photos you see in the chapter on engine block reconditioning.

While photographing at Evans I met Eric Hanson of Stage V Engineering, Corp. Eric builds the most beautiful wedge to hemi conversions you've ever seen. It was his 440 block the folks at Evans were machining. If you want one of these fabulous engines, get in line now—Eric has customers in line, worldwide!

For top Mopar performance, who's the best? Dick Landy Industries, of course. My sincerest thanks go out to Dick for reading parts of the manuscript, thereby keeping me from making a couple of serious mistakes. Robert Landy, Dick's son, was always available for help and information. In fact, Robert introduced me to Roland Osborne. This is how the hobby keeps going and progresses—we all help one another. Big thanks also go to Mike Landy, one of the best in the business.

When I needed photographs of crankshaft work, I looked for the best. I found it at Velasco's Billet Crankshafts in Downey, California. Henry Velasco and Chris Bynum gave me free run of the shop to take my photographs. Putting up with a photographer means stopping every couple of minutes while the picture is being taken. Thanks, Henry and Chris, for taking the time.

When you're writing a book about Chrysler, you can't do it without the help of Chrysler. I'd like to thank Larry Shepard of Mopar Performance plus Dale White and Jim Stoneman of Westlake Chrysler/Dodge who man the Mopar Performance desk there.

While I'm thanking parts people I'd like to also thank Carl Carr and Tim Mints of Carr's Auto Parts in Escondido, California. Carl and Tim are long-time friends. This is the second book they've helped with.

Other Escondido friends who contributed immeasurably are all of the great people at the

Mattman Company: Bill Mattman, John Mattman, Kim Wheeler, Gary Jackson, Glen Bishop and Bob Lane.

When I needed something, a part, a tool, a helping hand or anything, my friend Jerry Olmsted, "The Birdman Of Escondido," was there as he has been through all of the books I've written. Now, his son, Rod, and daughter, Julie, are old enough to be mentioned, as they helped, too.

Having helped even longer—since he was old enough to hold a wrench—is my "kid" brother, Al Taylor. No longer a kid now, but a full-on automotive engineer, Al's help was but one more time in thousands.

At The Corvette Shop, Larry Hofer, Tim Turner and Caroline Davidson teamed up to help. Larry is an automotive genius, a walking encyclopedia of Chevy, Ford and Mopar knowledge. Caroline and Tim kept things going while Larry set me straight answering questions for hours on the phone when he should have been working.

Jake Erlandson of Ace Transmission in Alpine, California, helped immeasurably with photos and information.

Thanks also to David Duncan of Chrysler Performance West, Galen Govier of GTS, Tom Race of RT Specialties and Julius Steuer of J&C Import Car Care.

Finally, I'd like to thank the man who took the basic manuscript, read it, corrected the mistakes, tweaked it, tuned it, adjusted it and made it into a book—Michael Lutfy, Automotive Editor Extraordinaire at HPBooks. Thanks Mike, for making me look so good! ■

PARTS, MATERIALS & INFORMATION

Chrysler Power Magazine
2044 Central Ave.
Duarte, CA 91010
(818) 303-6220

RT Specialties
501 W. 8th Ave. #5
Mesa, AZ 85210
(602) 827-0171
**Source for Mopar shop manuals*

Automotive Book Stop
1508 W. Magnolia Blvd.
Burbank, CA 91505
(818) 845-1202
**Source for Mopar shop manuals and reference material*

Evans Speed Equipment
2550 Seaman St.
So. El Monte, CA
(818) 444-2838
**Custom engine machining*

Stage V Engineering Corp.
P.O. Box 827
Walnut, CA 91788
(909) 594-8383
**Race-bred hemi heads for Wedge blocks*

Dick Landy Industries
19743 Bahama St.
Northridge, CA 91324
(818) 341-4143
**Builders of high-performance Mopar engines*

Mopar Performance Technical Assistance
P.O. Box 215020
Auburn Hills, MI 48321
(313) 853-7290

GTS
Galen Govier
Rt. 1, Box 322K
Prairie Du Chien, WI 53821
(608) 326-6346
**Specialty books on Mopar casting/forging numbers, VIN numbers and other documentation*

Velasco Billet Crankshafts
12422 Benedict Ave.
(310) 862-3110
**Custom crankshaft grinding*

Bob Mazzolini Racing
3199 Columbia
Riverside, CA 92591
(909) 787-8783
**Specializing in Max Wedge racing engines and parts*

INTRODUCTION

The story of Chrysler's big-block engines is the story of their entry into the world of brute horsepower and high performance. It is the story of one horsepower for every cubic inch of displacement; a milestone in the auto industry similar to exceeding the speed of sound in a piston-engined aircraft. The glory days of Chrysler racing are also a part of this story. So are passenger cars equipped with engines powerful enough to drive off the showroom floor and onto the drag strip. Names like Keith Black, Dick Landy and Tony Nancy help write part of the story on the nation's race track.

The history of the big-block Chrysler engine begins in 1958. This is the year the "B" series engines were introduced in the 350 and 351 cubic-inch versions. It took until 1963 for these engines to mature, but when they did they became some of the highest performing engines available. The basic "B" block grew from its modest beginnings as a 361 into a 383 and 400. All three displacements had a stroke of 3.38-in. and respective bores of 4.12, 4.25 and 4.34-in.

In 1962, the most powerful 383 to hit the street came equipped with dual 4-barrel inline carburetors, special "300" series cylinder heads and a hot camshaft. The factory rating for this smallest of the big blocks was a whopping 345 horsepower.

Meanwhile, Chrysler engineers were looking for more torque—and that meant a longer stroke. Thus, in 1959 a long stroke version of the "B" block was introduced. This was called the raised block or "RB" engine. The block casting was modified (or raised) to increase the cylinder bore length by 0.745-in. This allowed the engineers to increase the stroke from 3.38-in. to 3.75-in. The result was the 413 cid. The 413 powered Chrysler's first high-performance "B" car, the Chrysler 300E.

This engine was the "RB" 413 with dual 4-barrels on a long-ram manifold. Compression was 10:1 with a factory horsepower rating of 380 at 5000 rpm with 450 lbs-ft. of torque at 3600 rpm. However, the biggest changes were yet to come in the form of the Max Wedge. This was the most radical departure from standard engineering design that Chrysler had offered to date.

The Max Wedge was brute horsepower. My first "new" car was a 1962 Plymouth Fury with the Golden Commando option—the Max Wedge engine. This 413 produced 420 horsepower. At 60 mph, you could step down into passing gear and the rear tires would still "burn rubber," at least until Officer Brown caught you doing it—twice. The 426 was introduced in the Chrysler that same year.

A new block casting was made to incorporate larger oil feed passages and to accommodate the big 1-7/8-in. exhaust valves. To do this, notches were fitted into the cylinder bores. The block was also stress-relieved. Forged TRW pistons were attached to beefier connecting rods. The compression ratio was 13.5:1.

Special cylinder heads wre developed that breathed more freely and accommodated the larger valves. With two Carter carburetors and in race tune, this engine developed 425 horsepower at 5600 rpm. In 1964, Chrysler reached its zenith with the introduction of the Stage III Max Wedge. This very high performance engine was, and still is, one of the most powerful high performance engines ever to be offered to the public at large. In 1963 and 1964, the 426 Max Wedge could be bought for less than a thousand dollars more than a small-block "A" engine. But time, tide and fast cars roll on and Chrysler rolled with them.

At the end of 1964, the Hemispherical Chambered 426 cid (Hemi) engine burst onto the muscle car scene. Although the Hemi had 426 inches of displacement like its Wedge counterpart, the similarity ended here. This engine reintroduced the hemispherical-shaped compustion chamber and increased intake valves to 2.25-in. and the exhaust to 1.94-in. The block was recast to accommodate the changes in the head and was beefed up to absorb the increased horsepower. With a single carburetor, Chrysler gave the engine a horsepower rating of 400. With dual-quads, they quoted 415 horsepower and the 12.5:1 compression version was rated at 425 horsepower.

My research indicates these were deflated figures from Chrysler, given at the time to confuse the competition. *Hot Rod* magazine, in their 1965 article "Modern Hemi from Chrysler," states that these figures, corrected to SAE standards, would be about 550, 560 and 570 horsepower respectively!

Chrysler's love of brute force would bring one more engine into the horsepower wars—the "RB" 440. It incorporated a carryover head from the Chrysler 300 series. This utilized huge valves with excellent breathing capabilities. In 1967, it powered the Plymouth GTX and Dodge Super Bee. The 440 heads proved to be real champs and were used to give one last hurrah for the 383.

In 1968, the 383 was equipped with 440 heads, a high-rise 4-barrel intake manifold and the 440 high performance camshaft. These were available in both the Road Runner and Super Bee.

Chrysler would make one final lunge at the gold ring before pollution controls and gas shortages ended the era of the muscle car. In 1969, the factory introduced the 440 "Six-Pack"—three 2-barrel carburetors drawing in 1200 cubic feet of air per minute! These beauties were produced through 1971 and ended Chrysler's 20-year streak of producing some of the best high performance street cars the world has ever known.

SHOP CLASS

The way I figure it, you purchased this book for one or all of the following reasons: you're interested in Chrysler big-block engines; you're going to overhaul your big-block engine; or you liked the cover. I hope I can help you with the first two reasons. I didn't take the cover shot so I can't claim credit for it.

I'd like to make a couple of suggestions to you if you bought the book for the first two reasons. Read the book first from cover to cover. There are a lot of little tricks you can pick up or learn. If you're going to rebuild your engine, I strongly suggest you read the book first, then read each chapter as that area comes up before jumping in.

The idea behind all of this is that it will acquaint you with what is going to happen, and you'll be able to prepare for it. This will save you untold hours of frustration and a few dollars to boot. The idea is to be totally prepared; to have all of your parts chosen and purchased, to gather all of the necessary tools, and to have your machine shop picked out. There is nothing more frustrating than getting under the car only to find you haven't what you need for the job. I've seen more than one project bite the dust because the rebuilder just wasn't prepared.

One by one, the chapters will tell you how to remove the engine, disassemble it, inspect and recondition the parts, assemble it and finally, slide it back into the engine compartment. The last part of the installation chapter includes a sidebar on tune-ups and break-in.

But there's a long way to go before you need to start worrying about breaking in and tuning in a new engine. I've tried to make that journey as safe, as quick and as fun as possible. I've designed this book with the hope that it is the only source you'll need to completely rebuild your big-block. Professional and shade-tree mechanics, as well as Mopar enthusiasts from all walks of life, have contributed to this book, and I hope you'll benefit from their collective knowledge. Good luck with your project, and may the garage gods smile favorably upon you. ■

INSPECTION

1

So it's time to do something about that engine of yours. It's burning some oil, it pings even on the best gas you can buy and it can't get out of its own way! It's do something now or send it to the crusher. Well, hold off on the crusher a minute and let's look at your options.

One of the most important things my dad taught me was to always look first for the simplest solution to a problem. If the car won't start in the morning, see if you have gas in the tank before you call the tow truck. Pretty simple, but I've seen people replace a starter when the solenoid was bad, replace carburetors when the gasket was leaking and get a full valve job when all that was needed were new valve-stem seals.

In this chapter we'll look for some of the simpler solutions to the problems of

Larry Hofer is listening to the heartbeat of the head assembly. He has placed the probe of his stethoscope near the area he suspects a problem may be. Note that the probe is not on top of the valve cover. The air inside is not a good conductor of sound.

excessive fuel and oil consumption, loss of performance or engine noises. Maybe you can squeeze another 30,000 miles out of the old beast before you have to do a full-on overhaul.

EXCESSIVE OIL CONSUMPTION

When it seems like you're putting a quart of oil in the engine every day you should begin to think it's time to do something about the oil consumption.

What is excessive? If your engine is using a quart in less than a thousand miles, something is very wrong. If it's using a quart in less than 500 miles, something is seriously wrong!

Oil consumption—and low oil pressure—are caused by increased clearances, which allow excess oil to pass. Assuming the oil is not leaking out on the garage floor, the most likely causes of excessive oil consumption are: broken or worn rings, worn valves or valve guides or deteriorated valve-stem seals. All are indicated by the same thing—smoke at the tailpipe.

PISTON RINGS

Three rings per piston seal the piston in the bore; two compression rings and one oil-control ring. The two top rings seal compression and combustion pressures while the bottom, or oil ring, controls the bore lubrication. The oil-control ring allows some oil to pass through to lubricate the cylinder walls and rings. When it wears, excess oil passes into the combustion chamber and is burned.

The loss of sealing also works the other way. Combustion gases escape past the compression ring and into the crankcase. This is called *blowby*.

Before 1966 (1964 in California) blowby forced crankcase vapors into the atmosphere through a vent in the crankcase. After 1966 blowby was vented into the intake manifold through the PCV (Positive Crankcase Ventilation) system. It was thereby drawn into the combustion chambers and burned with the fuel charge.

Excessive blowby can actually pressurize the crankcase of an engine and cause oil to weep from every gasket. When this happens, there may be no single, severe oil leak, but the engine and engine compartment may be covered with oil.

Valve Guides

Worn intake valve stems and guides (or worn rings) allow excess oil into the combustion chamber. This happens when the car slows down and speeds up again, and a big cloud of smoke comes belching out the tailpipe. During acceleration, high manifold vacuum pulls oil down the valve stems and past the rings. When you accelerate again, the oil mixes with the fuel charge and blows smoke everywhere. Of course, the problem is aggravated by worn or cracked valve-stem seals, if any are even left. Gases may also escape through the worn exhaust valves and guides into the rocker arm area and then into the crankcase.

Sometimes, when you first start the engine, great billowing clouds of smoke will appear. This is usually the result of stem-and-guide wear. Oil seeps into the combustion chambers while the engine is off. When you start the engine—stand back! Although you will have oil passing through these worn valve guides, you should also understand that some loss occurs even when everything is new. Some oil must be consumed through the normal lubrication process.

Sludge buildup in the head can compound oil consumption through the valve guides. Sludge can restrict the oil drains in the head and prevent oil from draining back into the crankcase. The head and valve cover then fill with oil and oil drains down the valve stems. This is particularly common in poorly maintained engines, or engines with solid lifters because the oil drains in those heads are fairly small.

Solutions—If you're suffering from high oil consumption—more than a quart every two thousand miles or so—but power is good and the rest of the engine checks out, you might consider installing new valve-stem seals. There's a nifty trick to do this without removing the heads, which is shown on page 112.

POOR PERFORMANCE

Two problems contribute to poor performance: increased fuel consumption and power loss. Fuel economy is measured in miles per gallon. When this drops you know you've got something wrong. Power problems are more elusive. As your engine wears through age, the loss of power is not readily felt because it has been so gradual. It can, however, be measured on a dynamometer. If you notice that your car is a bit slower accelerating from a stoplight, or sluggish in the hills, then performance has deteriorated.

If you feel the engine is performing below normal but oil consumption is within limits, try a thorough tune-up. Remember: start with the easiest solution first. It must, however, be a first-class, major tune-up. Take the car to a reputable shop with a dynamometer. With this they can measure horsepower at the wheels. By measuring the engine output before and after the tune-up, you can determine to what extent the tune-up corrected the problem.

Causes of Poor Performance

The biggest cause of poor performance—lost power and economy—is worn piston rings and cylinder bores. Increased blowby, accompanied by the loss of compression, causes an engine to perform at less than peak efficiency. This is usually the main cause of the problem.

If oil consumption is okay but mileage and power are down, look to the valvetrain. If carburetion and ignition are working correctly, check for blown head gaskets, carbon buildup, burned exhaust valves or worn camshaft lobes and lifter.

Blown Head Gaskets—There are two ways a head gasket usually blows out: between two cylinders, or between a cylinder and a waterjacket. If the gasket is blown between cylinders, compression will be down in both cylinders. A blow-out between one cylinder and a waterjacket will affect only that one cylinder. On page 7 you'll see how to run a compression test.

There's a quick and simple test to see if a head gasket is leaking into the cooling system. Start the engine with the radiator cap removed. Never remove the radiator cap from a hot engine. The pressurized coolant will come blowing out and scald you for sure!

If there is a leak into the cooling system, you'll see bubbles rising to the coolant surface. These bubbles will be full of exhaust gas. As they pop, little clouds of vapor will be released. If you're not sure what you're seeing is exhaust gases, sniff them. If you smell exhaust, you have exhaust. If you want to be completely sure, take your car to the shop. They'll have a mechanical "sniffer" that can do the job. Unfortunately, this doesn't tell you which cylinder is leaking. This takes a compression or leak-down test which I describe on page 7.

Burned Exhaust Valve—A burned exhaust valve cannot seal, consequently, the cylinder with the burnt valve will be down on compression and power. Burned exhaust valves are more common on engines with air injection systems. This is the result of continued combustion behind the valve in the exhaust port and manifold.

Worn Camshaft Lobes & Lifters—Camshaft lobes and lifters wear together. When this wear is excessive the valves don't fully open. With an intake valve the result is reduced fuel charge, which produces insufficient combustion pressure.

A worn exhaust lobe can cause the same problem. The reduced valve opening prevents all of the burned fuel charge from escaping the cylinder, leaving less room for a new fuel charge. The result is the same: reduced power. The procedure for checking for worn lobes and lifters is shown on page 10-11.

Carbon Deposits—Carbon deposits are the result of how the car has been driven. In areas where most of the driving is done on turnpikes and freeways, cars seldom develop these deposits unless there is severe carburetion problems. The big culprit is stop-and-go driving under 30 mph for extended periods of time. These deposits can fool you into thinking the engine needs to be rebuilt.

When carbon builds up inside an engine's combustion chamber, it takes up room, increasing the compression ratio. This can lead to *detonation.* Detonation is caused by the fuel charge exploding—detonating—when pressure from normal combustion causes spontaneous ignition of the unburned fuel charge. Another problem, *preignition,* results from the hot carbon particles igniting the air/fuel mixture before the spark plug does. Although the causes are different, both preignition and detonation are often referred to as pinging. It's that irritating rattle you may hear when the engine is under heavy load and you're using a low octane fuel.

Sometimes the engine will continue to run, albeit erratic, after the ignition is turned off. We call this *dieseling,* though the better description is post-ignition. Here, carbon deposits heat to the point that they begin to act like glow plugs in a diesel or model-airplane engine. This causes the engine to continue to try to run even though the ignition is off. Dieseling can also be caused by too high an idle speed. So, don't assume your engine has carbon deposits just because your car diesels.

Carbon buildup is the result of short-haul drives, rich fuel mixture, oil getting into the combustion chamber or long periods of idling. If you cure these problems the carbon will go away. If, however, it's a serious problem, look to the chapter on cylinder-head reconditioning and remove the carbon manually.

Low Compression—Wear and tear on an engine produces a lowering of compression. The two top rings of the piston that seal the compression will wear, or even break. Cylinder bores also wear. Generally if the wear is uniform throughout all eight cylinders the only thing affected is acceleration and top speed. If one or more cylinders suffer more wear than their counterparts, severe power loss can occur. Erratic running and extreme mileage loss can be the results. In the next section I discuss problem diagnosis. The section on compression testing will tell you all about your engine's compression.

DIAGNOSIS

Now that I've scared you with all the possible problems your engine could have, let's take a look at how they can be diagnosed without tearing into the engine. Even if your engine shows no specific problem, you should perform the following diagnostics to answer our original question: Should the engine be rebuilt?

Internal Noises

So you have noises coming from the engine. These can tell you a lot of information. First, however, you must discover whether the noise is at engine speed or half engine speed.

Connect your timing light to the engine and watch it blink. If the noise is in time with the light, it's at half engine speed. If the noise is at double the time to the light, it's at engine speed.

If the noise is at engine speed—once for every revolution of the crank—the problem is probably in the bottom end, everything from the top of the block down. This could mean a broken piston ring, worn connecting-rod bearing, or a worn pin bore in a piston. There is one noise in the bottom end that occurs at half engine speed: piston slap. It occurs only on the power stroke.

Should the noise occur at half engine speed, it's probably in the valvetrain. Valvetrain noise can be excessive lash (if your engine has solid lifters) or a collapsed hydraulic lifter. A bent pushrod is another possibility.

On page 1, my friend Larry Hofer is seen listening to the engine with a mechanic's stethoscope. This is a great tool for listening to the internal workings of an engine. If you don't have one or can't borrow one, a long screwdriver or ratchet extension will work. Put the blade of the screwdriver or end of the extension on something solid, near where you suspect the noise is coming from. Touch the "listening device" to your head, just behind your ear. You'll hear all kinds of interesting noises.

Noises at Engine Speed—If you want to listen to the valvetrain, place your listening device on a head bolt. If you put it on the valve cover you won't be able to distinguish one noise from the other. This is because the airspace inside the cover will damp the noise to a point where you won't be able to distinguish where it comes from.

Now what are you listening for? What should you hear? From the bottom end, a broken piston ring makes a chattering or rattling noise. If you accelerate the engine, it will become amplified.

Piston slap is a dull or hollow sound. The piston is actually wobbling around inside the bore. A collapsed piston skirt causes a similar but much louder noise. This can best be heard when the engine is cold. It decreases as the engine warms up and the piston expands to reduce piston-to-bore clearance. To check for piston slap, retard the spark. Loosen the distributor lock-down bolt and rotate the distributor a few degrees counter-clockwise. If you're suffering piston slap, this should reduce the noise. Don't forget to reset the distributor after this test.

A light knocking or pounding noise that's not related to detonation or preignition is probably due to excess connecting rod bearing-to-journal clearance. This means the bearing is worn out. Finally, a light tapping or clicking noise may indicate excess pin-bore clearance in a piston.

To confirm and pinpoint a particular bottom-end noise, disconnect and reconnect the spark plugs one at a time. When you disconnect the wire from the bad cylinder, that cylinder is unloaded. Power is eliminated with the cylinder and the piston-and-rod assembly "relaxes," eliminating or reducing the noise it was making. The noise should disappear or be greatly reduced. If this test is being performed on a car with a solid-state (electronic) ignition, be sure the plug wire is grounded. If not, you run the risk of damaging the electronic control module.

Kits are available that make this a simple task. They consist of jumpers, or short wire springs, that snap over the plug end and into the plug wires. The cylinder can then be disabled by touching the spring with an insulated wire grounded to the engine.

Noises at Half Engine Speed—If you're hearing a sharp clacking or rapping noise at half engine speed, this indicates your engine probably has a collapsed hydraulic lifter. If the noise is a light clicking, it's probably excess clearance in one of the valve mechanisms. If this clicking gradually diminishes or disappears after the engine is started in the morning, for example, the problem is a lifter with excessive leak down. Due to wear, oil leaks out of the lifter and it collapses under the load of the valve spring. When the lifter is pumped up—pressurized with oil—the noise goes away, but a completely failed lifter will not pump up.

To double-check a noisy valve lifter,

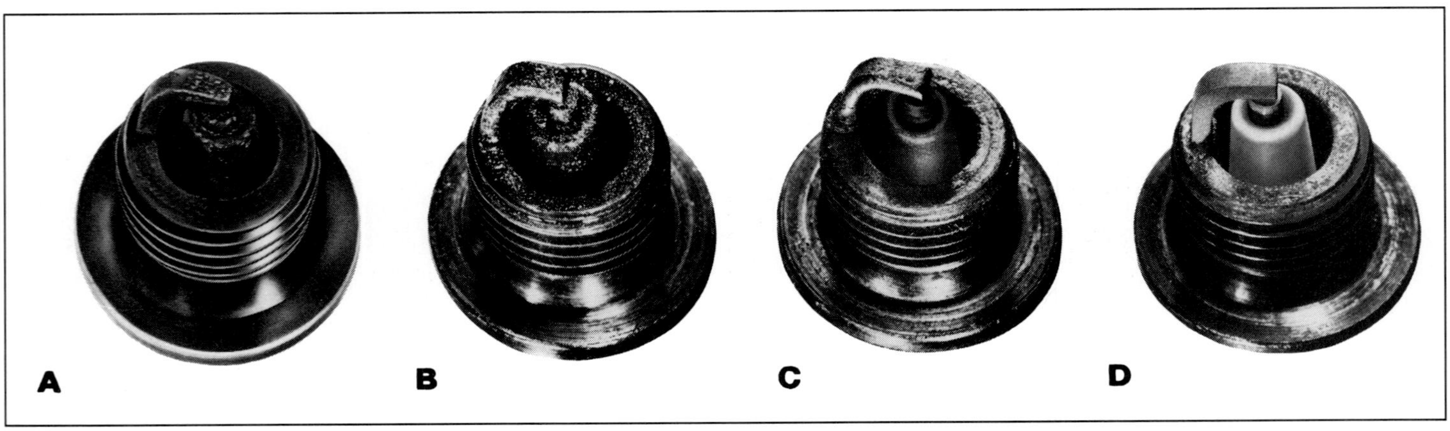

"Reading" spark plugs is a good diagnostic technique. Worn-out plugs, A, are easily recognized by eroded electrodes and pitted insulator. Replace them and your engine's performance will instantly improve. Oil-coated and fouled plugs, B, indicate internal wear of piston rings, cylinder bores and valves guides. Carbon-fouled plug, C, has dry, black and fluffy deposits, and indicates weak ignition, carburetor problems or poor driving habits. Normal plug, D, has brown to grayish-tan appearance with some electrode wear indicated by rounded edges on center electrode. When checking plugs, record the cylinder each one came from. ***Photos courtesy Champion Spark Plug Company.***

remove the valve cover on the noisy side. With the engine idling, insert a feeler gauge between each rocker arm and its valve stem one at a time. When you get to the offending valve, the noise will stop or change pitch considerably because the feeler gauge has taken up the clearance. Suspect a malfunctioning lifter or bent pushrod, particularly if the engine has been over-revved.

I assume your engine is not equipped with solid lifters, which by design click all the time. If the engine has solid lifters and is extremely noisy, a simple valve adjustment may solve the problem.

Checking Spark Plugs

Inspecting each individual spark plug will tell you a great deal about what's going on inside your engine. As you take each one out to look at its "innards," be sure you do not mix them up.

Wet, black deposits on the inside of the threaded portion of the plug and on the plug insulator are caused by excessive oil passing the rings or intake valve guides. Oil lost through exhaust valve guides doesn't enter the combustion chamber, so deposits won't show on the plugs.

If the plugs show black, dry, fluffy carbon deposits, one of the following is to blame: over-rich fuel mixture, wrong spark-plug heat range, weak ignition or sustained low-speed driving at high engine load. All the plugs should look alike if any of these problems exist. If carburetion is at fault, the carburetor must be checked. Normally, plug fouling from an over-rich mixture indicates a fairly severe carburetor problem like flooding, internal leakage or a sticking choke. These will show up first as a drastic drop in fuel economy and black exhaust smoke. Plugs fouled from driving habits or wrong heat range should be replaced by hotter plugs.

As friend Larry continues his diagnosis, he connects a vacuum gauge to the intake manifold. The gauge is holding steady, indicating all cylinders are drawing equally. With this same vacuum gauge Larry can also create a vacuum using the hand pump.

Using both types of compression gauges to check compression. Retesting after squirting about a teaspoon of oil into the cylinder will determine if this weak cylinder is a result of poor rings or valves.

Consider taking the car out for a long freeway or turnpike drive twice each month.

TESTING

There are a number of instrument tests you can make to tell you some very sophisticated things about the workings of the engine. These include: cranking vacuum test, power-balance test, compression test and leak-down test. Let's look at each of these to see what they tell us.

Cranking Vacuum Test

The cranking vacuum test shows how well the cylinders seal in relation to one another. To perform this test you'll need a vacuum gauge. Connect the gauge to the intake manifold and start the engine. If the gauge indicates a steady vacuum after it stabilizes, all eight cylinders are sealing the same in relation to one another. If the needle fluctuates, in a pulsating vacuum, one or more of the cylinders has a problem. It could mean one of several things: poor valve timing due to incorrect adjustment, a worn camshaft lobe or collapsed valve lifter, a leaking valve, worn cylinder bore or piston rings or a leaking head gasket. The cranking vacuum test tells us there's a problem. The other tests tell us where it is.

Power Balance Test

To perform the power balance test you'll need a very accurate tachometer—not the type that comes stock on your car, but one that reads in single rpm. You'll need one accurate enough to determine a 20 rpm differential.

This test determines whether all of the cylinders are working equally. The method is to "kill" each cylinder, one at a time, by grounding the spark plug and comparing the rpm drop. This indicates how much each cylinder is contributing to the overall power of the engine. The less the speed drops, the less that cylinder contributes.

Install the spark plug jumpers or disconnect all the leads at the spark plugs and reinstall them loosely. With an electronic ignition, don't disconnect the plug wires; use the grounding technique.

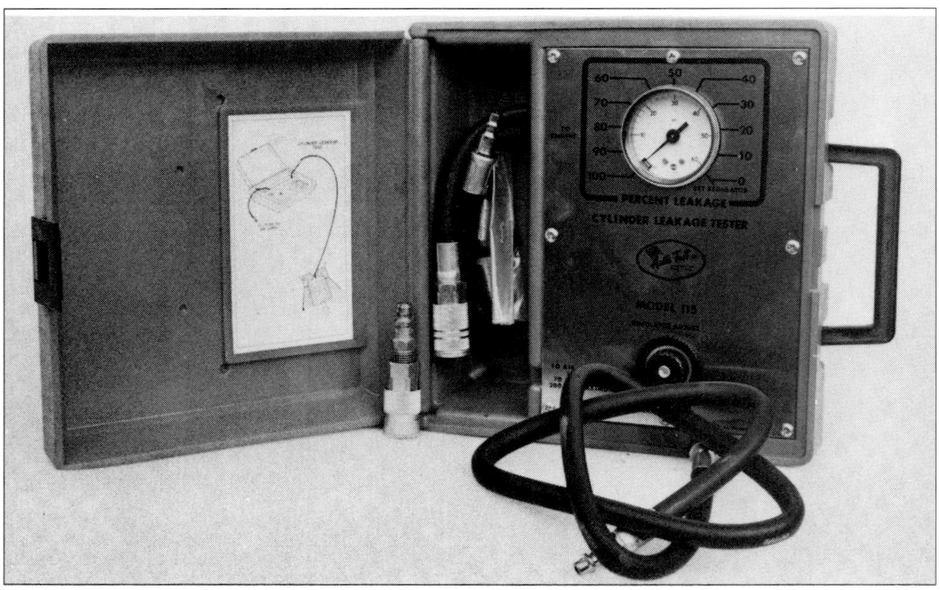

A leak-down tester is the most accurate way to measure a cylinder's sealing capabilities, because it eliminates factors that can affect compression testing. Leak-down test is done without cranking the engine, so cranking speed, valve timing and cam condition do not affect the readings.

With the engine warmed and running and the tachometer connected, set the idle speed at about 1000 rpm. Ground or remove the number 1 spark plug lead. When engine speed stabilizes, record the drop in rpm. With the number 1 cylinder firing again, repeat the process with the next cylinder. It's probably easiest to go down one side—1, 3, 5, 7—then move to the other side checking 2, 4, 6 and 8. It doesn't make much difference as long as you record the rpm drop and indicate the cylinder number.

When all eight cylinders have been tested, compare the rpm drops. If all are within 20 rpm, no one cylinder is much worse than any other. But, if one cylinder is different by as much as 40 rpm, that cylinder has a problem. Note the ones with the least drop, and pay particular attention to them during the compression test.

Compression Testing

Compression testing is the best overall test for engine condition. Often, it's the only test done if the mechanic is in a hurry. This test give an overall comparison of the cylinders, but doesn't give the answers a cylinder leak-down test can. Leak-down testing is discussed next.

Like the power balance test, you need three things: note pad, remote starter switch and a compression gauge. The screw-in type gauge gives the surest reading. The gauge with the rubber cone is less expensive. I recommend the screw-in type as it is virtually foolproof. The rubber cone type can slip and give an inaccurate reading.

Run the engine until it warms up to normal operating temperature. As the exhaust manifolds are very hot, be careful as you perform the next step. Remove all the spark plugs. Prop the throttle open and be sure the choke is open. Insert the compression gauge in the number 1 spark plug hole. Crank the engine over about five times and make a note of the reading. It's important to crank the engine the same number of times for each cylinder. If you turn it over 5 times on number 1, then be sure you turn it 5 times on number 2, 3, 4 and the rest. When all eight cylinders have been checked and duly noted, the test is complete. So now what? Well, depending on your engine, its condition and problems, you'll have a wide variety of readings. Our 383 cid project car should range between 125 and 155 psi; a 426 Hemi will be in the 150 to 204 psi range and 440 buffs will get readings from 130 to 165. Obviously, these are very wide ranges. The rule is, to hope that all of the cylinders are within 75% of one another. For example, if the highest reading cylinder shows 160 psi, the others should not be less than 120 psi (160 X .75 = 120).

Usually, if you get a 0 to 30 psi reading, you can figure a bad valve. Poor sealing rings will usually give pressures over 75 psi. In the above example of a 120 to 160 psi range, should any cylinder fall below that range, there's a problem. To determine the cause, squirt about a teaspoon of oil into the weak cylinder. Squirt it toward the far side of the piston so the oil will run down between the piston and cylinder wall all the way around. Give the oil a few seconds to run down, then check the compression again. If the reading increases, the ring and bore are at fault. If the pressure doesn't change, you've got other problems. To check those out, look to the leak-down test.

Leak-Down Testing

Leak-down testers are rather expensive. If you have access to one, fine; beg or borrow. If you don't have one I'd suggest taking the car into a tune-up shop for this test.

The leak-down test is similar to a compression test, except the pressure is created externally and you measure the rate at which it leaks down. This test eliminates many of the variables that affect compression testing: valve timing,

VALVE JOBS ON HIGH-MILEAGE VEHICLES

Making an arbitrary decision to "get a valve job done" on a high-mileage engine can lead to a host of problems. If you take the heads off, send them to a machine shop and have them rebuilt, the result may be worse oil consumption and blowby than you had before.

The reason for this is how an engine wears. All the parts wear at about the same rate. But the reconditioned head seals better, creating higher compression and vacuum loads on the rings and pistons. Where the rings and pistons were doing a moderately good job in a moderately good engine, they may not be able to seal against the higher compression. The result is increased oil consumption and more blowby.

camshaft wear and engine cranking speed.

This is a fairly simple test. Be sure the piston is at top dead center (TDC) of its power stroke. To determine TDC of the power stroke—both valves closed—trace the spark plug wire from the cylinder being tested back to the distributor cap. Make a mark on the distributor housing just below the cap where the plug wire is connected. Then, remove the cap. By turning the engine over until the rotor is directly over your mark, you can be sure the piston is near TDC on its power stroke.

When you perform this test, an air-hose nipple is threaded into the spark plug hole and the cylinder is pressurized. The pressure is read on a gauge that indicates the percent of air leakage from the cylinder. A good cylinder should not lose more than 5 to 10 percent. If the loss exceeds 20 percent, something's wrong.

When you're doing a cylinder leak-down test, remember the air that's escaping has to go somewhere. There are only five places the air can go: the intake manifold, exhaust manifold, crankcase, another cylinder or into the cooling system. Where it goes will tell you a lot about what's wrong.

Rings—With the cylinder pressurized, remove the oil-filler cap and listen for escaping air. If the rings are bad enough, the pressure will force oil right out the dipstick tube.

Head Gasket—If air is escaping between adjacent cylinders, you can bet on a blown head gasket. Remove the spark plugs from both cylinders. If you hear air from the spark plug hole of one while the other is pressurized, you've probably got a blown (burned) head gasket. For full confirmation do the valve leakage check that follows.

A blown head gasket into the cooling system will only affect one cylinder. To check for this problem, remove the radiator cap. Fill the radiator to the top and watch for large bubbles. If the

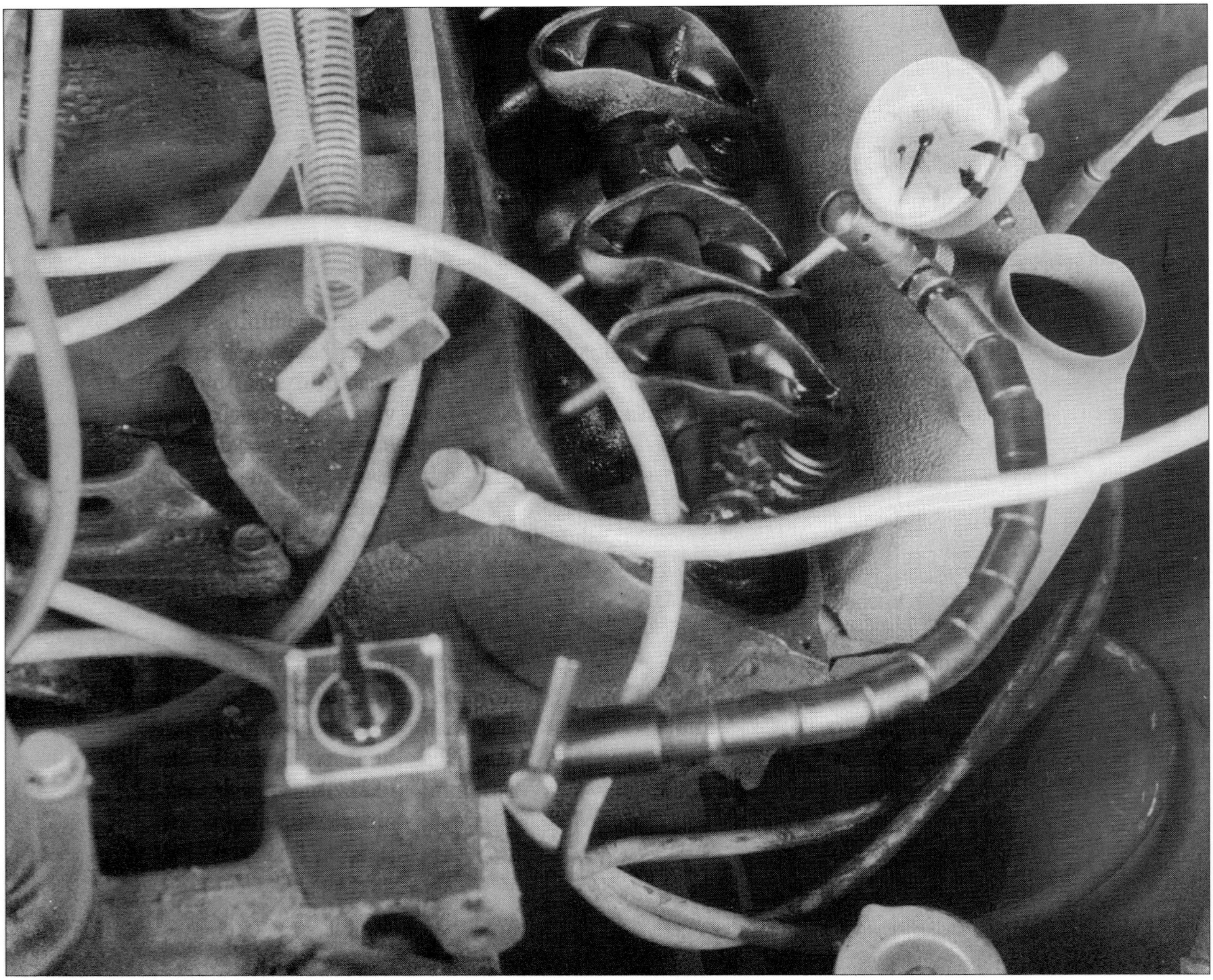

The dial indicator will give you the most accurate reading when measuring camshaft lobe lift.

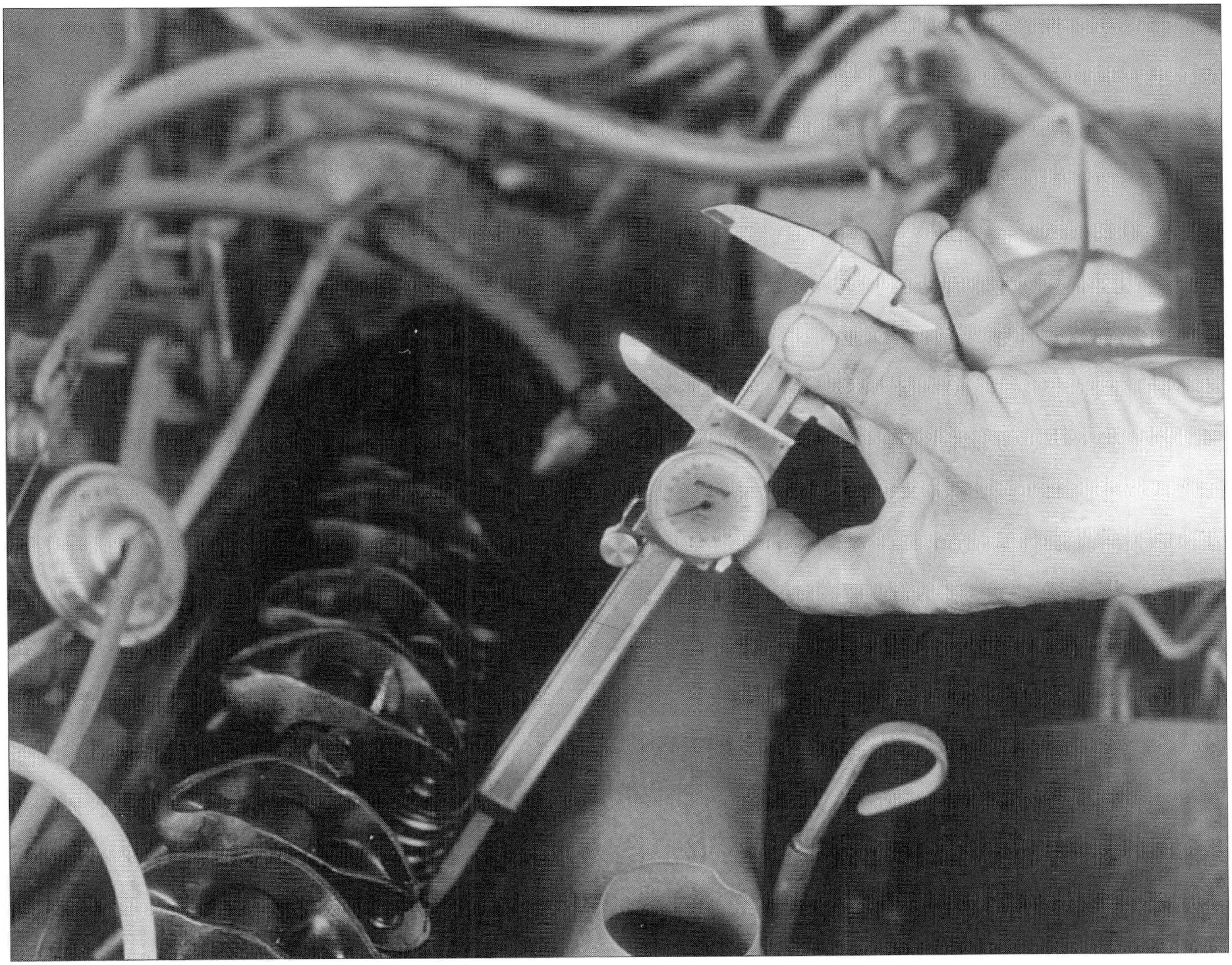

If you don't have a dial indicator, using a vernier caliper to check the difference in valve height between open and closed will give you a fairly accurate picture of the cam lobes.

problem is a blown head gasket, install a new one. You'll find these instructions given in the chapters on teardown and assembly. Often, a blown head gasket is the symptom of another problem: overheating, a warped head, or from a burn-through. Be sure you repair whatever caused the problem in the first place or you'll be doing the job all over again.

Valves—With the cylinder pressurized, listen for escaping air at the carburetor and tailpipe. Air escaping through the carburetor indicates a bad intake valve; air out of the tailpipe means a bad exhaust valve. Poor seating is the reason for valve leakage. It could be a burned valve, broken spring, insufficient valve lash or even a bent valve stem.

For most of these problems you'll have to remove the head—that is unless the valve spring has broken or the engine has solid lifters and the valve has insufficient lash.

If the engine has adjustable rockers, check valve adjustment (see page 136). If adjusting the valves corrects the problem you just saved a lot of work. If the problem is a broken spring you may also have a bent valve stem as the result of the piston hitting the valve head. You can replace a broken spring without pulling the head, but the head must be removed to repair a bent valve.

CHECKING THE VALVETRAIN

A valve that doesn't open fully will not affect cylinder sealing but will seriously affect power output. Therefore, if one of the cylinders fails the power balance test but passes the compression test, check the valvetrain.

Remove the valve cover on the affected head. Before checking the valvetrain take a close look at the head. With the valve cover off you can get an indication of the general condition of the engine. Heavy, carbonized oil deposits indicate poor maintenance (infrequent oil changes) or overheating. If you find metal filings you're getting poor lubrication—often the result of clogged oil passages. Sometimes you will find bits and pieces of valve-stem seals. High engine temperatures harden the seals and they crack and break up. You'll also find these pieces in the oil-pump pick up screen in the oil pan.

Broken Valve Spring

If you find a broken valve spring, replace it the same way you replace valve-stem seals (page 112). While the spring is off, reinstall the spring retainer and check for installed height. This is explained on page 105. This will tell you if the valve is closing fully. If the valve won't close, the head must come off to repair it.

Checking Valve Opening

With the valve cover off and a piece of cardboard propped up on the head to deflect oil, start the engine. While it's running, look closely at each rocker arm and valve. If a valve is not opening fully, chances are you'll spot it. The rocker or valve will move very little compared to the others. If you find one that appears to have problems, check it.

The're are four reasons a valve won't open fully: a worn rocker arm and shaft, bent pushrod, collapsed lifter or a worn camshaft lobe.

Rocker Arm Failure—Rocker arm failure is more common on engines with stamped-steel rockers than those with cast or forged rockers. Stamped rockers are only surface-hardened. Once this hard surface is worn the arm wears rapidly. Often, you'll find a small hole worn through a stamped rocker arm where it meets the pushrod. To check further for worn rocker arms look for metal filings around them. They may also be loose on the shaft compared to the others.

As you crank the engine, the rocker arm will appear to be working properly. However, the arm must move up to contact the shaft before it opens the valve. Depending on how badly it is worn, the arm will actually open the valve only slightly.

The following checks require it, so remove the rocker shaft now if you need to look further. To prevent bending the shaft, remove the bolts by turning each one turn at a time, to gradually unload the bar from the torque stress. Then remove it completely. Rocker arm failure is usually the result of clogged oil passages. Check these carefully.

Bent Pushrods—If the rocker arm checks out okay, remove the pushrod and check it. Pushrods are the weak link in the valvetrain. Bent pushrods are usually caused by over-revving the engine. Occasionally a foreign object will jam in the valve spring and prevent the valve from opening fully. This will also bend the pushrod. Usually it's obvious to the naked eye if the pushrod is bent. If you're not sure, just roll it across a flat surface and you'll see.

Collapsed Hydraulic Lifter—The easiest way to find a collapsed lifter is by process of elimination. If the rocker arm, pushrod and camshaft lobe lift check out, all that's left is the possibility of a collapsed lifter. To double-check your reasoning, reassemble the valvetrain but leave the valve cover off. Start the engine. Place a finger on the valve end of the rocker arm. If the lifter has

OEM CAMSHAFT LOBE LIFT

Originally Installed Camshaft

Engine	Intake At Valve	Intake At Lifter	Exhaust At Valve	Exhaust At Lifter	Intake Opens Before TDC
383	0.425	0.283	0.435	0.290	16°
400	0.434	0.289	0.430	0.287	18°
413	0.389	0.259	0.389	0.259	14°
426 Hemi	0.467	0.311	0.473	0.315	30°
440	0.450	0.300	0.458	0.305	19°

collapsed, you'll feel a distinct shock wave as the valve opens and closes.

A collapsed lifter will also be noisier than the others. You'll hear a distinct "clack, clack" sound. If you insert a 0.010-in. feeler gauge between the valve stem and the rocker, the noise should lessen as the feeler gauge takes up the slack. Lifters can be replaced without rebuilding the engine. You'll have to remove the intake manifold, however. Discussion of intake manifold removal begins on page 37.

Camshaft Lobe Wear

Camshaft lobe wear does not show up in a leak-down test or compression test. It's a physical measurement and requires one of two measuring devices: a dial indicator as first choice or a vernier caliper as second. The most accurate is the dial indicator.

Mount the dial indicator, using the magnetic base, so it is lined up with the center of the pushrod. Push on the pushrod to make sure it is well seated in its lifter. Slowly bump the engine over with the starter while you watch the indicator. When the pushrod reaches its lowest point, zero the indicator. (You do this by rotating the face of the dial until the "0" is aligned with the indicator mark at the top of the dial.) Now bump the engine over again. Note the maximum indicated reading and record it. Indicate whether it is an intake or exhaust valve and the cylinder number. You'll need this information later for comparison to the other lobes. Camshaft lobe lift is the difference between the highest portion of the lobe and the base circle diameter. Lifts will vary depending on engine type. Refer to the specification chart below.

When checking camshaft lobe wear, you'll know right off which one is bad. Rather than reading in thousandths of an inch, the difference will read in tenths! Generally, lobes wear out one at a time. It would not be unusual for one lobe to be totally worn while the others remain in good condition.

Although I'm demonstrating on a small-block, the idea is the same. At top I'm using vernier calipers to check valve height between open and closed. But the most accurate method is to use a dial indicator *(bottom photo)* to measure lobe lift directly from the pushrod.

If you choose to use a vernier caliper, check the difference in valve height between open and closed position. This will give you a fairly accurate picture of the lobe lift.

Regardless of how many lobes you find worn, it only takes one to require cam replacement. When you replace the cam you must replace the lifters too. Otherwise, the new cam will be ruined before you get the car out of the driveway. Use the procedure outlined in the engine assembly chapter when changing a camshaft and lifters.

Replace the cam and lifters if they're bad. But, if they're bad and you have excessive oil consumption plus compression in one or more cylinders below the 75% mark, it's time to rebuild. ■

ENGINE REMOVAL 2

In Chapter 1, you tested, checked and probed to determine whether or not "Betsy" had a few thousand miles left in her. But everything you found pointed to the inevitable conclusion that Betsy was either going to become an artificial reef for the fish or she was going to have to be completely rebuilt. Unfortunately, the law prohibits you from dumping a car into the bay so you've selected the second alternative—an engine overhaul.

In the following chapters, we'll take it step-by-step. I'll discuss safety, what you'll need for preparation, how to do each step of the process and show you all the tricks I know to make the work simple and easy. Read through the chapter first so you'll know what to expect. There's nothing worse than digging into a job and finding halfway through that you don't have the needed tool or part or, worse yet, you realize you need a friend at 10:00 PM to help you with some procedure. But first, the safety lecture.

SAFETY

If you know the rules, please read the following as a refresher. If this is your first serious exposure to engine work, pay very close attention. It may save a few scraped knuckles, a finger or even your life! Everyone has heard at least one story of the car that fell on a neighbor or relative. Somewhere it really happened.

When working with an automobile engine, you're working with 600-700 lbs. If this weight were applied to your foot, it

Now and then you'll need a helping hand. Here's the "now." Exercise extreme caution because the hood is heavy. Make sure you make locating marks by tracing the hinges around the hood with a grease pencil or permanent marker. Otherwise, realignment later will be tough.

would be extremely painful. If it dropped from just a few inches above your foot, it would crush it. An engine dropped from a height of four feet would cut your foot off and crack the concrete below!

For example, probably the worst, and unfortunately, most common way to lift an engine is by slinging a chain hoist over one of the roof beams of the garage. Don't do it. More than one garage roof has ended up on the garage floor!

Today, you have access to more sophisticated equipment, easily obtained and quite inexpensive. You can rent a device called a *cherry picker*—a hydraulic hoist. Casters allow the hoist to be rolled into place. When the engine is pulled from the car, it can be rolled away. I demonstrate this on page 25. A set of large, fixed wheels on rental models allows the unit to be towed to your work site with a car or truck. The cherry picker provides the best and safest way to remove an engine.

Fender covers are a great item. They keep the fender from being scratched, especially by belt buckles, and some have ridges to keep tools from rolling away.

Next and equally important: Never get under a car without safety stands or jack stands. Any time a car is jacked up, the first things under the car should be the safety stands. Place stands under a frame member or unibody side rail with the car resting firmly on the stands. Don't trust your life to a hydraulic jack or concrete blocks. If the jack should blow a seal or get kicked while you're under the car, and there aren't any safety stands, chances are you're done for. Masonry blocks can't stand single-point loads and can easily crumble.

These are lifesaving safety precautions. Other injuries can be avoided by using common sense. Wear safety glasses if there is any possibility of flying dirt or metal. Use them anytime you're under the car. Use a box-end or socket wrench rather than an open-end wrench whenever practical. Pull on the wrench rather than push. The two above suggestions will help eliminate skinned knuckles. Though not done often enough, I know two mechanics who wear light leather gloves when they work. This eliminates a few more of the

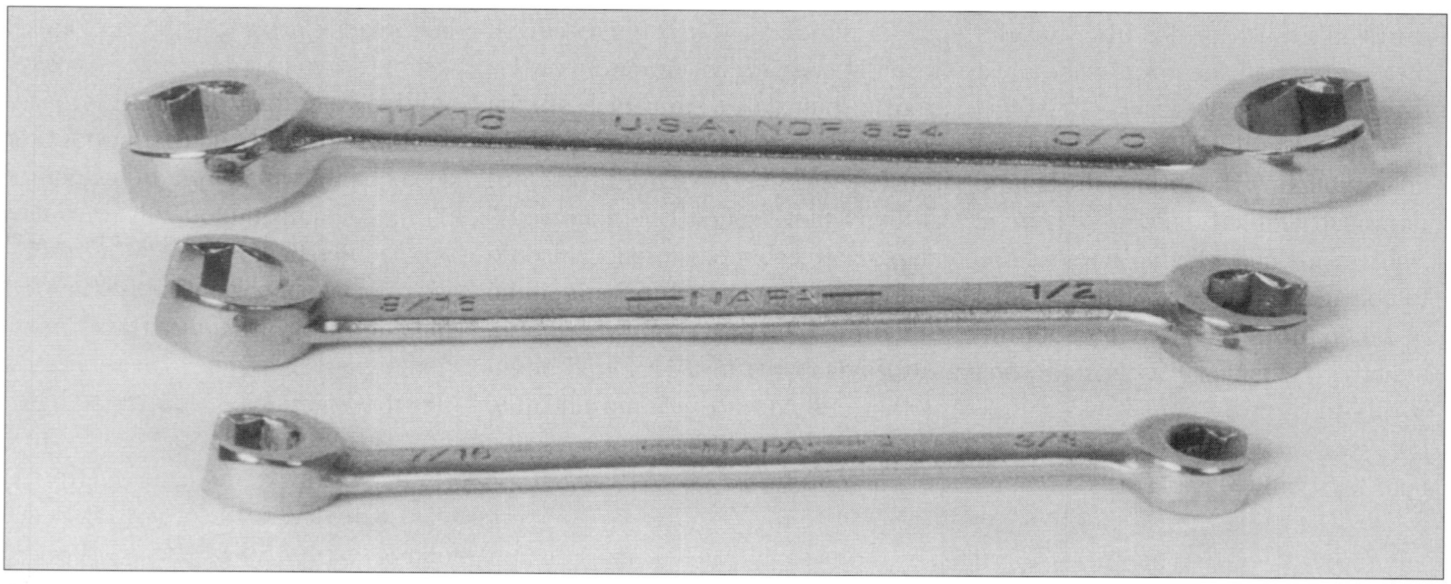

Save yourself a great deal of headache and money by investing in a set of flare-nut wrenches for removing fuel and transmission cooling lines. Unlike box-end and Crescent wrenches, they are designed to prevent rounding of the nuts. If you round a fuel line nut, you're looking at cutting and reflaring the tubing (possibly even replacing the entire line) to fix it.

EPA REGULATIONS

There are ways each of us can work to save our environment; and in the doing, save ourselves from possible confrontations with the authorities. It's against the law to dump engine coolant, engine oil, ATF or any other environmentally harmful substance anywhere but at EPA-approved receiving stations. There are now severe fines for dumping.

Each of these fluids must be captured and kept in separate containers, then taken away for proper disposal. If you mix any of them together, no one will take it.

Many parts houses are receiving stations for these products, as are auto repair facilities and some recycling centers. Call around on the phone before loading up and trying to find someone to take 5 quarts of used oil off your hands!

As of this writing the EPA was not forbidding you to bleed off the freon from your air conditioner. However, we all want to do what's best for the environment. Therefore, if there's an EPA certified air-conditioning service center in your area, why not take your car in and have them capture the freon. The price is minimal, your safety is better assured and the air you breath will be just a bit nicer. Remember, you have to find this same service center to have your system charged. You can no longer buy freon in the little cans and fill the system yourself as of August, 1992.

nicks, scrapes and burns and keeps your hands from getting roughed up and permanently embedded with grease.

Don't work on a hot engine. Never try to drain hot oil or coolant. Don't smoke while disconnecting fuel lines. If you have any appliances in the garage with open pilot lights, such as a hot water heater, either extinguish the pilot light or move the vehicle out of the garage before disconnecting any fuel lines. Like the car that fell on the friend or relative, we all know a story of the garage and house that burned down when the gas fumes hit the pilot light. All of these precautions may seem rudimentary, but for every one broken there's been an injury. Please don't become a statistic or one of the "percentages."

Finally, never rush your job. The best way to foul up is to hurry. When you hurry you get careless and when you're careless you get hurt and do the job wrong. Give yourself plenty of time. Set no unrealistic goals. Don't try to work all night. This isn't the Indy 500 and the race doesn't depend on you.

Okay, the lecture is over, class dismissed. Let's pull an engine!

PREPARATION

The key to any successful job is organization. Properly organized, your job will be easier and quicker. The bonus will come at installation time. This is true throughout the entire engine rebuilding process.

Tools

In addition to the minimum amount of mechanic's tools and equipment we've already discussed, you'll need masking tape, a pen or pencil, assorted cans or small boxes for nuts and bolts, drain pans for water and oil, a spray can of penetrating oil for stubborn bolts, degreaser and paint for marking things.

Fender covers will save fender paint and be a handy place to lay your tools. You'll add to your creature comforts with a creeper. A piece of heavy cardboard or Masonite will do if you're working on grass or dirt. Coat-hanger wire is indispensable to the home mechanic.

A Polaroid camera is a tremendous asset. Use it to photograph throttle linkage, clutch linkage, wiring connections, or anything you may need to remember how to put back together again. When engine installation time comes these photos will be a great assembly aid. The latest technology, and a piece I've added to my collection, is the miniature tape recorder. Like pen and pencil, you use it to make notes as you work—and you don't have to worry about spelling!

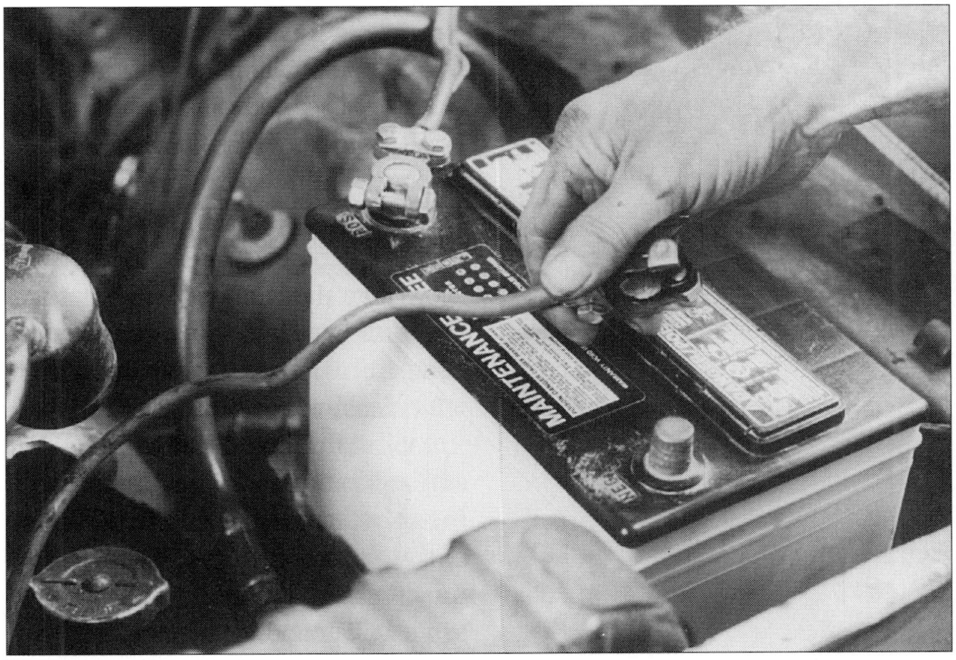

First step is to remove the battery. Disconnect the negative cable first. Note the absence of a battery frame here. A fast corner and the battery meets the fan blades. I'll install a new one before the engine goes back in.

Clean Your Engine

The last preparatory step is to clean or degrease the engine and transmission. A clean engine is much nicer and easier to work on. Many communities have a coin-operated car wash with one bay for degreasing engines. This is the fastest and most efficient way to degrease an engine. Besides, you leave the mess behind, not in your driveway. Be sure, however, the car wash really will allow this. Just ask the attendant.

The next best way to do the job is to have it professionally steam-cleaned. Of course, this will cost a bit and is not available in every community. If your community permits it and it does not violate pollution control ordinances or environmental practices, you can use one of the commercial water-soluble degreasing agents. Put one or two gallons of this degreaser in a three-gallon insect sprayer. With the long, curved wand you can easily reach any part of the engine, top or bottom. Follow this with a good strong blast of water from the garden hose and the engine will be clean enough to work on comfortably. It won't be clean enough to paint, but you'll be able to find bolt heads. If you use the family garden sprayer, be sure to clean it well afterwards. The leftover degreaser might kill your plants.

After removing the bolts that retain the shroud, lay it back over the fan to keep it out of your way while you remove the radiator bolts.

If you don't like the idea of using the family sprayer, degreaser can be purchased in aerosol cans and used in a similar fashion. Some final words on degreasing: all the grease and dirt have to go somewhere. Don't do your cleaning in the garage or on the lawn. Try to find an area where the runoff will not offend plants or people. Also, avoid getting moisture in the distributor and carburetor. Cover them with small plastic bags before hosing things down. Dry the spark plugs and plug wires before attempting to start the engine.

With tools, materials, a clean engine and a willing spirit, you're ready to start.

The demonstration car is a 1972 Plymouth Satellite with a 383 engine. If you're building one of the other series, you should have no trouble transferring the information from the 383 to your particular model.

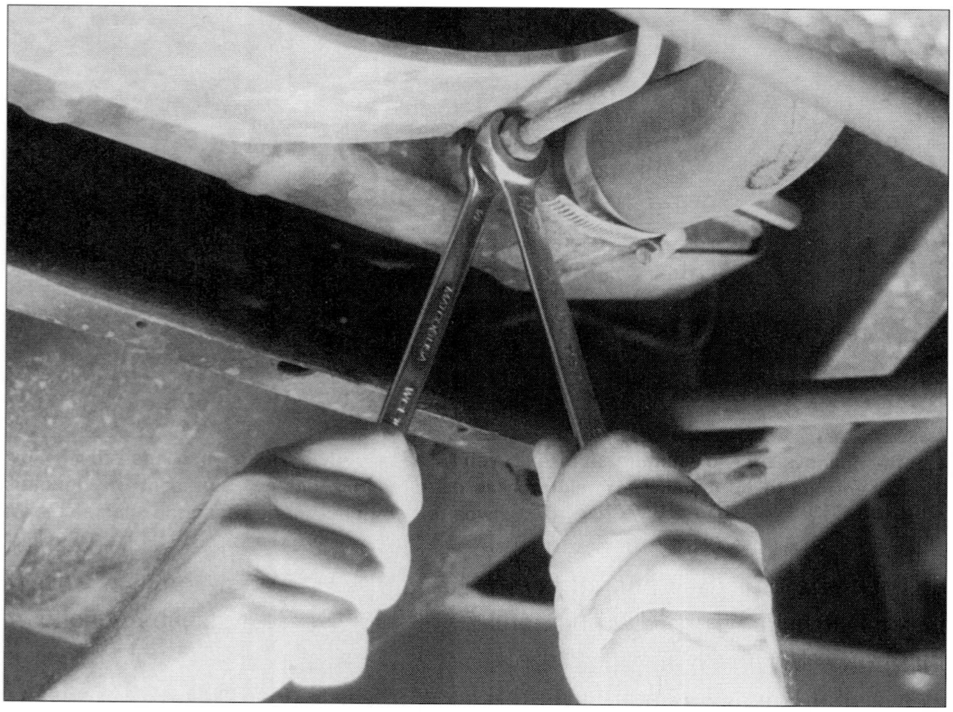

Note the hex to remove the transmission fluid cooling lines. This prevents twisting out the connector and rounding the shoulders of the flare nuts.

Hood

Hood removal is a two-man operation, so, enlist a friend for a few minutes. A hood is easily removed, but can take hours to align later if you don't have locating marks. This is easily done by transferring the hinge outlines to the hood with a grease pencil or permanent felt-tip marker. When you replace the hood, simply line up the hinges to these marks and the hood is aligned.

Another foolproof method is to drill a

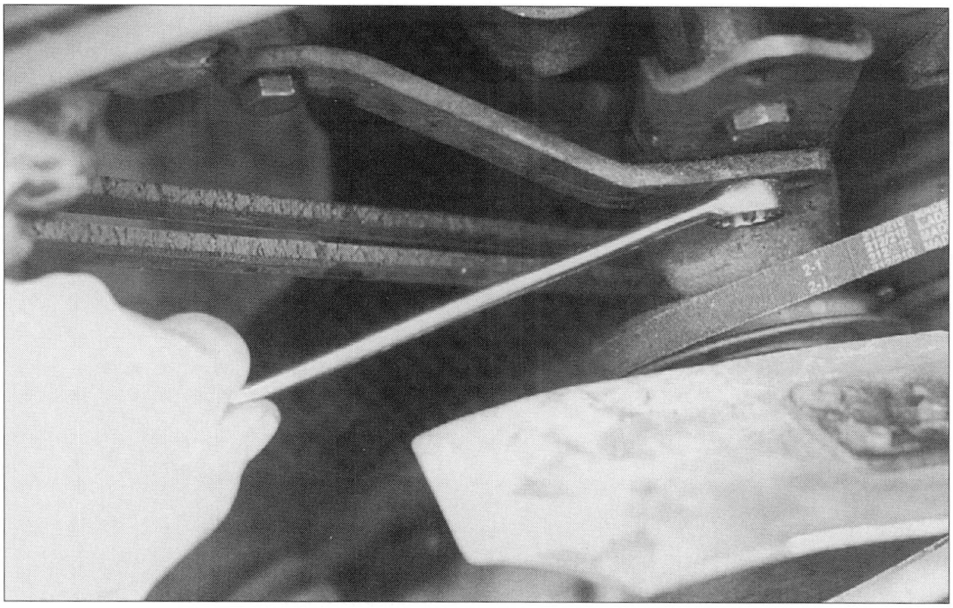

This is the water pump tension pulley. Remove the bottom bolt first, then loosen the top bolt. Now the pulley can be rotated to the side so the drive belt to the water pump can be removed.

1/8-in. hole through each hinge and hood inner panel. To align the hood, insert a nail, awl or drift pin through each hole and the hood will be in its original position. Rags taped to the rear corners of the hood will avoid chipping or scratching the paint on the hood corners or cowl.

Remove the front bolts first. The hood will pitch forward if you remove the rear bolts first. With a friend on one side and you on the other, remove the rear bolts and lift the hood off. Be careful! The hood is heavy and you'll be in an awkward position. If you're working outside, don't try to do this when it's windy. If the car is inside and won't be moved after the engine is removed, set the hood on the car's top with a blanket or large piece of cardboard underneath to protect the top (which should be set in place before you lift the hood!). Should you plan to move the car to another location, the hood can be stored in your garage, leaning against the wall. Wire the hood latch to a nail in one of the wall studs. This will keep it from being knocked over.

Battery

Remove the battery before you do anything else. This will prevent the possibility of any short circuits or shock. Remove the ground strap from the negative terminal first, then remove the strap from the engine block. If you remove the positive terminal first you run the risk of shorting something to ground as you work. Negative terminal first, then the positive.

When disconnecting the battery cable clamp, loosen the nut, then gently spread the clamp apart a small amount. Once it's broken loose from the terminal, rotate the clamp back and forth while pulling up. Don't attempt to pry it off with a screwdriver. This could pull the terminal out of the battery case.

Now remove the battery holder. Corrosion may have locked the nuts to the hold-down bolts. Here is where the penetrating oil comes in. A good soaking with WD-40, CRC, Liquid Wrench or Break-Free will have the nuts ready to loosen in minutes. As you remove the battery from the battery compartment be careful not to get it against you. The surface acid will eat holes in your shirt. Store the battery away in a cool, dry place where it won't be knocked over and spill the acid.

GETTING STARTED

There are no cut-and-dried engine removal rules, but plenty of tips. Simply proceed in what seems the easiest, most logical sequence. You're the judge as to which parts come off first. Normally, the best approach is to get the radiator and fan out of the way first so you can get to the accessories. This also gets the radiator out of harm's way.

Radiator

Drain the radiator and remove the hoses. To speed the draining process

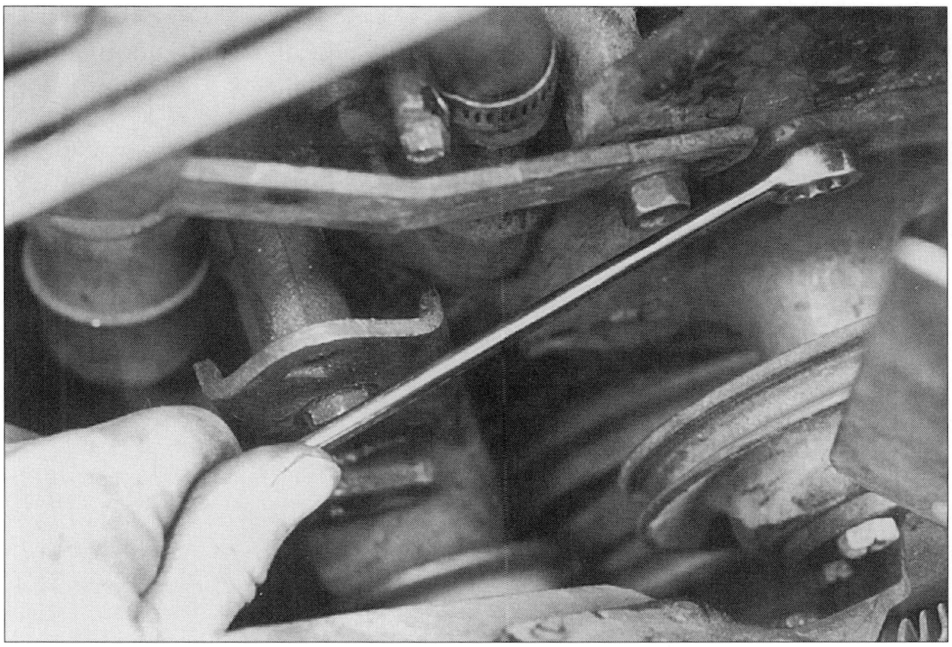

Remove the water pump and fan blades as a unit from the water pump housing.

When was the last time liquid passed through here? This corrosion is the result of not using coolant mixed with water in the cooling system.

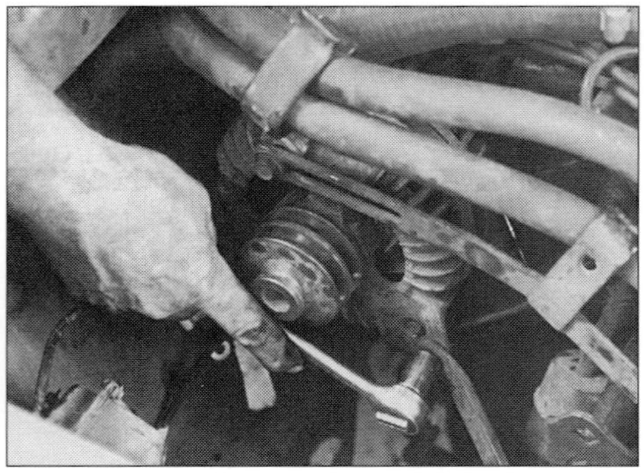

Loosen the top and bottom bolts to the alternator. This will release tension on the drive belts. Remove the top bolt last so you don't have to juggle the alternator around while you remove the bolt.

remove the radiator cap. This allows the radiator to "breathe." A petcock, or drain, is located at the bottom of the radiator opposite the radiator outlet. Place a large pan or bucket under the petcock and open it by rotating it counterclockwise as you view it head on. The petcock is soldered into the radiator tank and is easily twisted out with excessive force, so be careful. If it won't move with gentle pressure from a pair of pliers, apply some penetrating oil to the threads. If it still won't rotate, remove the bottom radiator hose. This will create a real mess, but at least the radiator won't be damaged.

To remove the hose, loosen the clamp at the neck and rotate it back and forth, breaking it free. Do not jam a screwdriver between the hose and neck to pry. The result will be a damaged neck and an expensive repair bill. Like all radiator parts, the neck is soft brass with soldered connections. If twisting won't free the hose, split it with a knife. If the hose is soft, stiff, cracked or more than two-years-old, it should be replaced anyway.

Begin marking things as you remove them for correct assembly later. I've marked these heater hoses "front" and "back," not the most professional nomenclature but nonetheless effective.

While the coolant drains, disconnect the top hose. When the radiator stops draining, disconnect the lower hose if you haven't done so. Watch out, it'll be full of coolant.

Radiator & Water Pump Removal—Remove the fan-shroud bolts, then lay the shroud back over the fan. If the car has an automatic transmission, disconnect the transmission cooler lines from the bottom radiator tank (see photo p.16). A flare-nut or line wrench will help prevent rounding the corners of the nut. When the lines are disconnected they'll leak automatic transmission fluid (ATF) so have your drain pan under them. ATF may siphon from the transmission through these lines. To prevent this, push a short piece of fuel line over the end of each tube, fold the hose back on itself and wire it in place. This will ensure the 13 quarts of ATF stay in the tranny.

To protect the fins, tape a piece of cardboard to the back of the radiator. Remove the radiator mounting bolts, then lift the radiator straight up and out. Be careful as the cooling fins are very delicate and easily bent if dragged across anything. While you have the radiator in your hands, check for signs of leaks—noted by rust stains—a bent or poorly repaired core, loose fins or any other signs of potential problems. Unless the radiator is in top condition, I suggest you have it reconditioned. It would be terrible to fry a brand-new engine because of a leaking radiator. With the radiator out, remove the thermostat housing and top radiator hose as a unit.

The water pump tension pulley is located to the left of the water pump as you face the engine. Loosen this by removing the alternator-support bracket then loosening the retaining bolt. Now you can remove the water pump drive belt. Remove the four bolts that retain the water pump to the block then lift out the pump and fan as a unit.

Alternator & Power Steering Pump

The next step is to remove the alternator. Begin with the wires. The top wire is the "battery" wire and is retained with a 3/8-in. nut. Below that is the "field" wire which can be pulled off by hand. There is no nut retaining it.

Begin now to mark things as you remove them. Make a masking tape flag for these two wires and mark them as "battery" and "field."

When you loosen the top bolt of the alternator you will relieve tension on the drive belts. They may then be removed. Leave this bolt in place and remove the bottom bolt. Go back and remove the top bolt with one hand while supporting the alternator with the other. Remove it and store it in a safe place. It will be easy now to remove the heater hoses. Be sure to "flag" them so they go back into the right place.

Use a flare-nut wrench here too, when disconnecting the high pressure ATF line of the power steering pump.

Turn your attention now to the power steering pump. It's located to the right of the air conditioner as you face the engine. You'll need a pair of hose-clamp pliers and a socket with a 1/2-in. drive.

With the hose-clamp pliers remove the hose at the bottom of the power steering pump. ATF is going to come pouring out so have your drain pan ready. When the line is clear the ATF will stop. You may want to plug the end of the hose or wrap a "small plastic bag" around it to prevent debris from getting in.

Disconnect the high pressure line (top). Use your flare-nut wrench on the flare-nut side and a regular box wrench on the hose side.

Where'd it go? There's supposed to be a line here from the air conditioning compressor to the condenser. We'll remove the A/C just as if it were there.

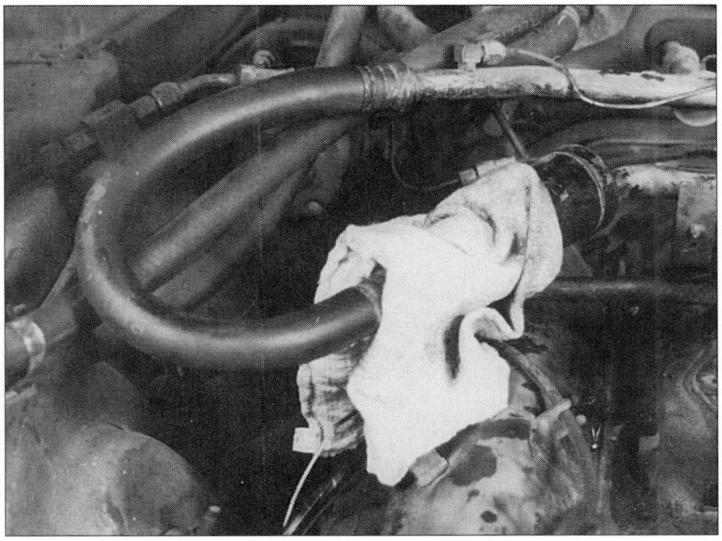

The shop towel around this broken connection helps contain the escaping freon. Without it the freon might have blown into my face as I broke the seal.

To prevent moisture and foreign objects from entering the A/C system, seal the open ends of the suction lines with masking or duct tape.

I recommend the 1/2-in. drive to remove the bolts retaining the P/S pump because they're on there to stay. It took quite a bit of horsing things around to get the bolts out on our project car. When you get it off, cover the intake and outlet to prevent strange things from getting inside.

Air Conditioning Compressor

You can see from the photo on page 19 that the project car's air conditioner had no freon in it as the line to the condenser was nonexistent. To demonstrate proper procedures, however, I went ahead as if it was all together and functioning.

Begin by releasing the freon charge (please see sidebar, page 15). Wear safety glasses and gloves while bleeding off the freon. The glasses protect your eyes and the gloves prevent frostbite—really! The freon is under great pressure. As it escapes and instantly changes from liquid to gas, there is a huge temperature drop.

Slip the wrench onto the connector. Then wrap a rag around the connector to contain the freon as it bleeds off. Carefully loosen the hose connection until you begin to hear a slight hissing sound, then stop. The sound is escaping freon—keep away from it. Once the freon has bled off—indicated by no further hissing—finish disconnecting the line.

After bleeding the freon, disconnect the other end of the steel suction line from the compressor. Be sure all open ends are taped off.

There are four bolts retaining the A/C compressor. Two are on the manifold, one of which is hidden under the coil. Remove the coil completely, and flag the wires attached to it. Now you can remove both bolts that secure the A/C compressor to the intake manifold.

The other two bolts are on the driver's side of the block. One also retains the

Here's the other end of the suction line. I've taped it and the compressor connection. You must do the same.

You could leave the steel suction line mounted to the engine; but we both know that's the best way to damage it as the engine is removed.

Beneath the coil is one of the bolts securing the A/C compressor bracket to the intake manifold. Remove the coil to access the bolt. Don't forget to mark the wires.

accelerator return spring bracket and the other secures the battery ground strap. When these bolts are removed the A/C compressor can be lifted out. Before putting it away be sure all the openings have been closed with tape. If you didn't do it first, remove the steel suction line.

Electrical

The top-side hard stuff is done. Now it's time to take care of the electrical components and wiring. Get out your pen or pencil and the masking tape and make a flag for every connector as you take it off.

A wire that's really easy to forget is the ground strap from the block to the firewall. Get this first. This grounds the engine to the firewall and subsequently to all your gauges in the dash. If you

There's that rascal. Now he's a cinch to remove.

neglect to replace it you'll get inaccurate instrument readings and other odd electrical disorders.

Using the molded boot, pull the wires from the spark plugs. As you remove them, flag them according to the cylinder number cast into the intake manifold near the valve.

Remove the distributor cap with all the wires intact and put it away. Now, disconnect and label the oil-pressure and water-temperature sending-unit wires. I disconnected the entire wiring harness to get it out of the way. I once tore one loose while lifting the engine and I don't want to do it again.

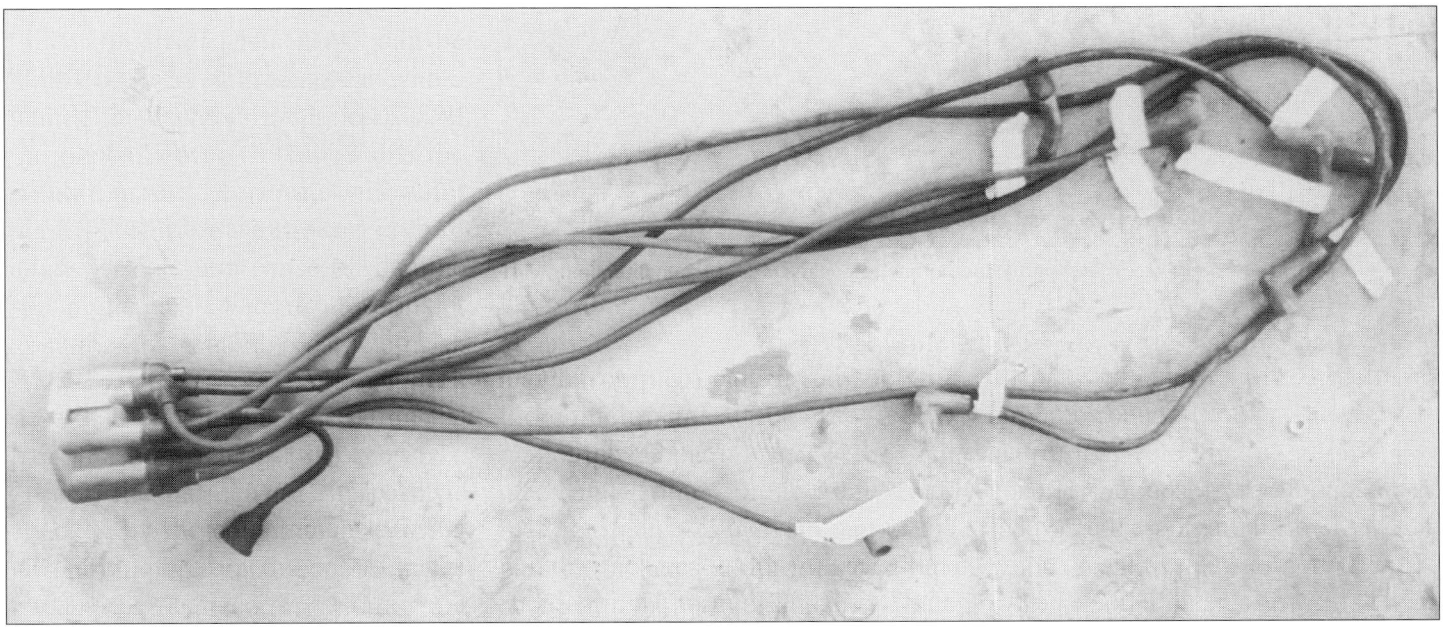

If you flagged each wire according to its firing order, you'll have no trouble lacing up the wires correctly when it comes time to assemble the engine. Forgot? Well, turn to page 155 and I'll show you how to do it "blind."

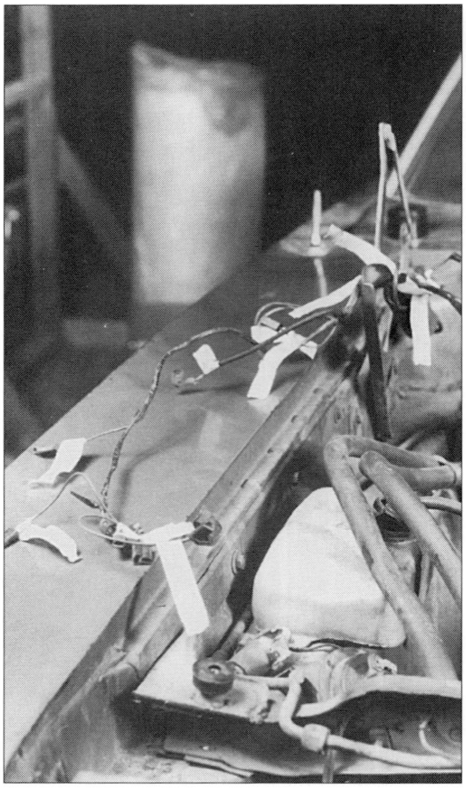

Here's the whole wiring harness tagged and out of the way.

I always put the clip back on the choke rod after disconnecting it from the choke lever.

Distributor—Disconnect the vacuum-advance hose at the distributor. A clamp and bolt secure the distributor to the block. Remove them and the distributor will lift out. It will take a little rotating to work it up and out—just don't pry or force it.

Throttle Linkage, Vacuum Lines & Fuel lines

In the safety remarks at the beginning of the chapter, I stated that all pilot lights should be extinguished before disconnecting fuel lines. Now is the time to do so! You're going to disconnect the fuel line at the carburetor and then at the fuel pump. Gasoline will drip out. The fumes will explode if they come in contact with flame. Remember: one cup of gasoline contains the same explosive potential as one stick of dynamite.

It isn't necessary to remove the carburetor prior to pulling the engine but it will ensure that it won't get damaged. To remove the carburetor, disconnect the fuel line using your flare-nut wrench. Remove the spring clip holding the automatic-choke lever in place and withdraw the lever. While you're on the passenger side, remove the PCV valve from the valve cover.

The throttle spring was probably removed when you pulled the air conditioning compressor. If your car had no air conditioning compressor, remove

After removing its clamp, slide the distributor out of the block. Rotate it a little as you go to disengage the drive.

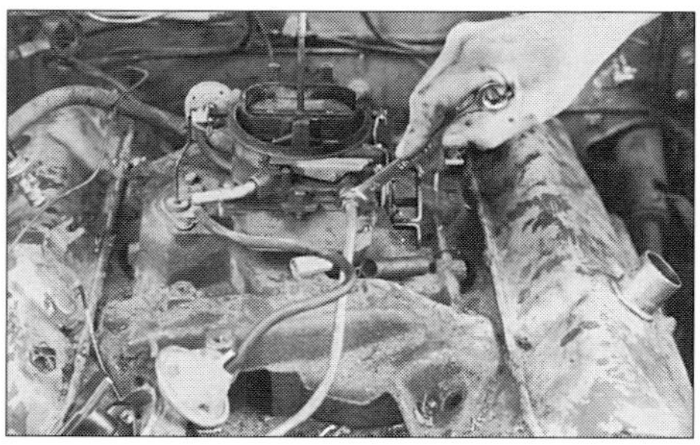

Another flare nut. Someone has always managed to use a Crescent wrench on these, rounding the corners. The design of the flare-nut wrench will help overcome this problem. Don't forget to turn off any flames or extinguish cigarettes while opening this line.

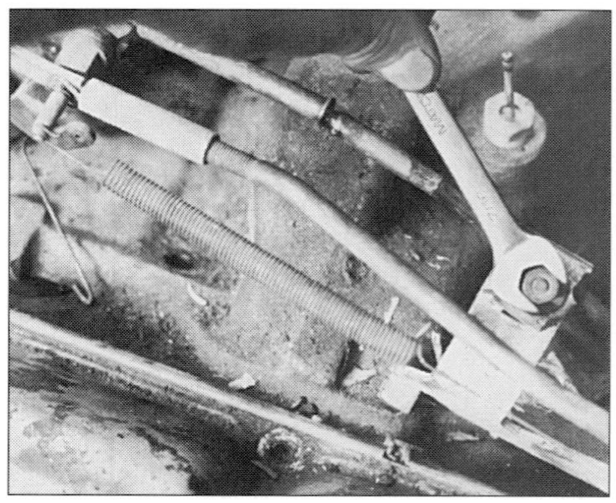

This clamp retains all of the firewall-to-carburetor linkage.

Yes, these photos are out of sequence. They were taken after the engine was removed. The photo on your left shows the location of the right (passenger) side engine mount through-bolt. It's a little hard to photograph while the engine is in. The photo on the right shows a 5/16-in. bolt sealing the fuel line.

the throttle spring now. Then, remove the kickdown-rod spring, the kickdown rod and throttle cable. To remove the throttle cable, loosen the clamp on the throttle-linkage support bracket. Remove the clip. Slide the connector off and the linkage should be free. Disconnect the kickdown lever at the rear of the kickdown fulcrum. This is an easy place to forget. If you do, when the engine is pulled you'll break the connection for sure. Two bolts and two nuts retain the carburetor to the intake manifold. Remove these and then remove the carburetor. Remember, it's full of gas. Don't tilt it.

Removing the carburetor enables you to use a lifting plate to remove the engine. This plate, which has a large U-bolt welded to it, fastens to the carburetor seat with the two bolts and two nuts you took out when removing the carburetor.

Finishing the Top Side

From the top you'll be able to remove the two top bolts that retain the bellhousing to the block. Take care here. The transmission dipstick tube is held in place by a bracket under the right bellhousing bolt. Avoid moving the tube while you're removing the bolt because only an O-ring seals the tube at the transmission. Moving the tube may break the O-ring seal, resulting in a transmission leak at the base of the dipstick tube.

Remove the top starter motor bolt. Go around to the right side of the car and remove the right-front motor mount through-bolt. The one for the left side will be removed from under the car. Disconnect the fuel line from the fuel pump by releasing the clamp and twisting the line back and forth. Again, gasoline will spill. After removing the hose, squeeze a 5/16- or 3/8-in. bolt into the open end and slide the clamp back up over the end, sealing it and preventing gasoline from siphoning from the tank.

This completes the top-side disassembly. You can now prepare the car to get underneath and finish disconnecting things down there.

UNDER THE CAR

Place the jack under one of the unibody side rails leaving room to place your safety stand. Jack the side of the car up as far as it will go and place your stand—carefully. Repeat this at the other side. Now, try to rock the car off the stands by pushing and pulling it. Better it falls now than when you're under it! When you feel secure that it's safe, get ready to remove the starter motor.

Starter Motor

Remove the starter motor cable and the starter solenoid wire. This wire is secured with a 11/32-in. nut. Don't confuse it with a 5/16-in. When the wires have been removed, remove the bottom nut. This is best done with an 8- or 10-in. extension on your 5/8-in. socket. The starter motor is very heavy, so be ready for it when you remove the nut.

Now is a good time to finish removing the bellhousing bolts. Leave one in to prevent the bellhousing from falling, so the flexplate doesn't get damaged

If you're working with a manual transmission, disconnect the bellhousing at the transmission, not the engine. Again, leave one of the bottom bolts in, this time to prevent damage to the pilot bushing and clutch. Disconnect the clutch-pedal pushrod at the bellcrank. Under the car, unhook the bellcrank-to-release-lever assembly. Be careful with all the clutch linkage hardware. The bushings, springs, seals and retainers aren't readily available.

Flexplate/Torque Converter

Remove the torque converter access panel by removing the four 7/16-in. bolts retaining it. One of the flexplate bolts is offset so it will fit the torque converter in one position only. Therefore, to save yourself an extraordinary headache during installation, apply some paint to one of the flexplate ears, bolt and torque converter. When you're ready to reassemble them you'll have a reference mark to index them in their original positions.

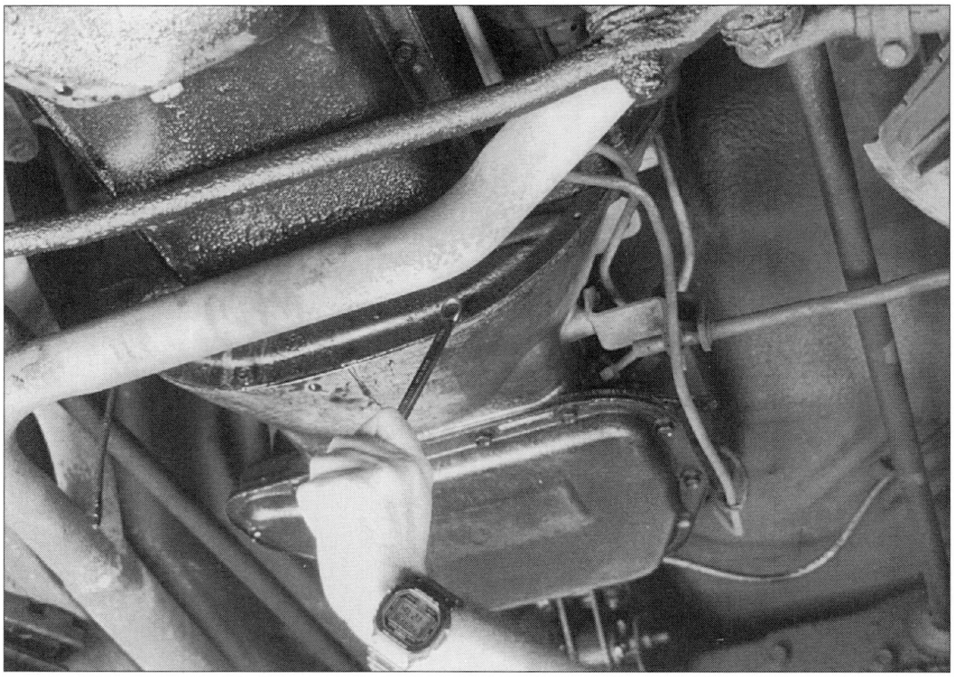

Using a box-end wrench to remove the flexplate access panel is a must. The heads of the bolts are very shallow and an open-end wrench will slip off.

The flexplate is secured to the torque converter with four bolts. Using a box-end wrench as a lever, place the box-end on one of the torque converter bolts and rotate the crankshaft clockwise—as viewed from the front—until the bolt is in its lowest position. The engine turns easier with the spark plugs out. Now, remove the bolt. Repeat this operation for the three remaining bolts. The torque converter will slide on the transmission input shaft, so push it back toward the transmission to prevent it from sliding off.

Exhaust Pipes

Here's one of the operations you don't want to start at 10:00 PM. You'll need a friend to help you out. He or she can stand on a small box and reach down with a long extension on a ratchet to hold the bolt while you work the nut from underneath. You'll need plenty of penetrating oil. Even with penetrating oil you may well twist the bolt in two. On the project car, I had to get on the nut with a breaker bar to loosen it.

Get out an aerosol can of paint and shoot some spray on one of the flexplate "ears" and its location on the torque converter. Match these up when you install the engine. It's the only way!

This is the proper location for your safety stands. If you locate them anywhere but under these side rails, you risk not only denting sheet metal but your body too!

Remove the left motor mount through-bolt. It's inserted through the motor mount in the opposite direction of the right.

Everything should be disconnected now. Check one more time and then jack the car up, remove the safety stands and lower the car to the floor.

ENGINE REMOVAL

The following operations must be strictly adhered to for safe and easy engine removal. If you're using a chain, it should be at least three-feet long. The chain should have at least 3/16-in. diameter welded links.

A chain that's too long can always be hooked up shorter—adjusted so the engine will clear the car before the cherry picker reaches its maximum height. The excess chain can be piled on the engine.

There is a good reason for the minimum-length chain. If the chain is too short, or hooked up too short, the engine could slide forward or backward through the hook due to the small angle of the chain. More important, the shallow angle places excessive loads on the chain, bolts and engine-attaching points. This greatly increases the danger of chain breakage and injury.

To avoid this problem, don't hook the chain up any shorter than necessary, certainly no shorter than three feet and use a fender washer between each bolt head and the chain. A fender washer is a large-diameter washer with a small hole.

Fasten the chain to the left front and right rear as you see in the photo. Make a final check for any wires, lines, hoses,

There's a reason for this photo. I once found a young fellow trying to lift out his engine with the cherry picker at the side of the car. What a time he was having! Run the hoist in under the front of the car until the arm is centered over the engine.

Keep looking for it, Mike. It's down there somewhere. This is what happens when you ask a painter in a white shirt to help you remove the exhaust-pipe bolts. Mike's going to find them located a little more towards his leg.

This cherry picker has an adjustable chain with a lifting block in the center. Note anchor points left-front and right-rear. Notice also that I've used fender washers between the head of the nut and the chain.

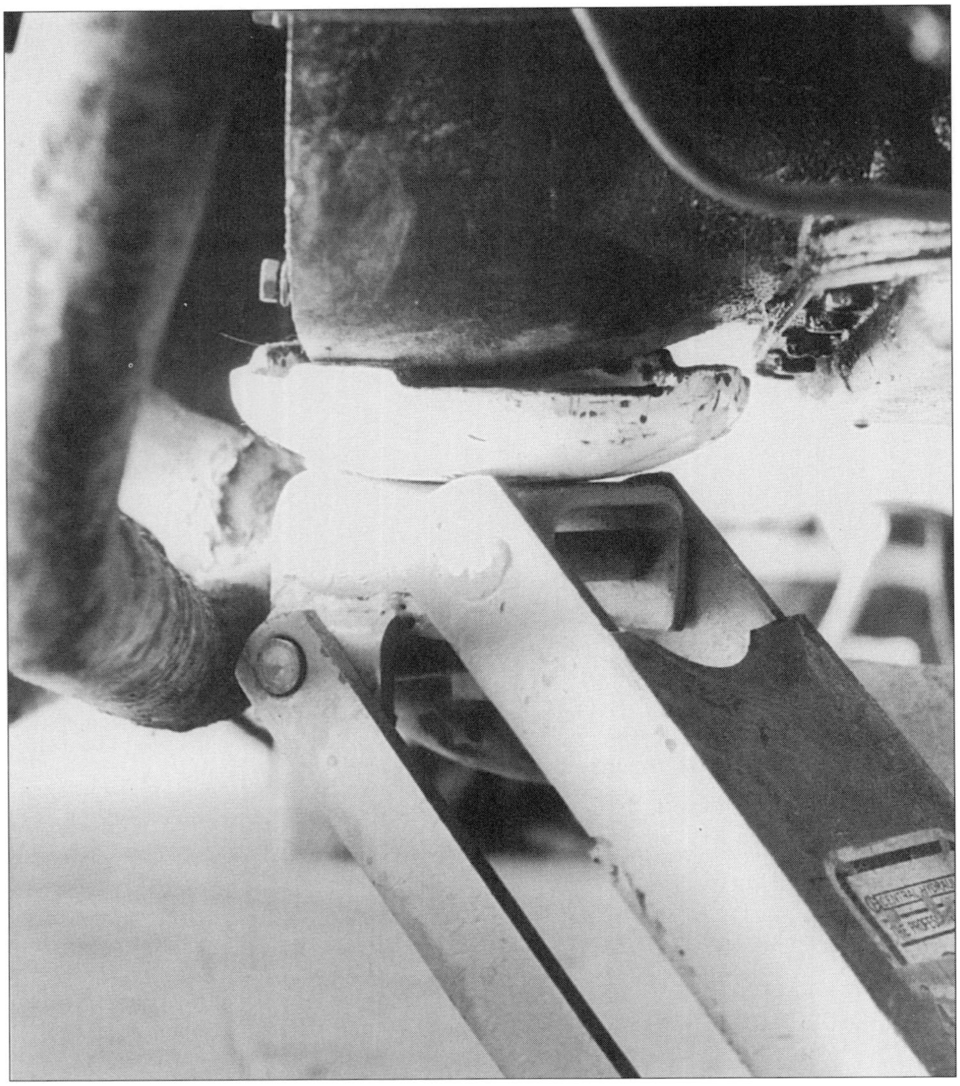

When the engine clears the engine mounts, place your jack under the bellhousing to support it. Now remove the bolt that was holding it to the block.

Here she comes!

bolts or other items you may have forgotten or overlooked. When you're sure everything is ready, position the cherry picker or hoist over the engine. Place the hook under the chain or through the lifting-plate loop and slowly start lifting the engine.

As the engine lifts, the front end of the car will rise with it. This is normal, because the front suspension is being unloaded. At this point, place the jack under the transmission and remove the remaining bellhousing bolt. Keep the jack under the transmission until the engine is out. This keeps the front end of the transmission from dropping and damaging the transmission mount, U-joints or shift linkage.

Raise the engine until the engine mounts clear the frame. Rock the engine from side-to-side to separate it from the bellhousing. If it won't separate, coax the engine and bellhousing apart with a large screwdriver or small prybar. Double check to make sure all the bolts are out first. When they separate, continue to raise the engine and pull forward. If the car is equipped with a manual transmission, the engine will have to come forward enough for the input shaft to clear the clutch.

Raise the engine high enough to clear the front of the car and wheel it out, or roll the car from under the engine. But before you move the car, tie the transmission up to the firewall. Old coat hangers are great for this. Now you can lower the car and remove the jack.

While the engine is suspended from the hoist or cherry picker, bolt the engine-stand plate to the rear of the block using the bellhousing bolt-holes. Place the engine stand onto the engine-stand plate and lower it and the engine as one assembly.

Congratulations! A job well done. You've finished removing the engine. In the next chapter we'll go through the process of disassembling it and getting it ready to go to the machine shop. ■

Before you remove the jack from under the bellhousing, wire the bellhousing to the firewall with some coat-hanger wire. Can you see the white paint on the torque converter?

While the engine is suspended from the cherry picker, mount the engine-stand plate on the rear of the engine. Lift the engine stand onto the plate, lock it in place and then lower the assembly to the floor.

All set for the next chapter on disassembly.

DISASSEMBLY 3

If you're this far into the book I can assume you got the engine out without too much trouble. Did you remember to congratulate yourself? Always pat yourself on the back for a job well done. It's something we all need; but there's never anyone around to do it for you when you need it most. Okay, with all the self-congratulations completed, let's get this engine block disassembled.

One of the first decisions you must make is what to do with all of the various nuts and bolts as you disassemble all pieces of the engine. You have two options. The first is to throw everything into a coffee can and send it to the hot (or cold) tank with everything else for cleaning. In this case, you'll have nice clean hardware with which to assemble your engine. Unfortunately, you won't have the foggiest idea of where anything goes! To eliminate this trial-and-error approach to assembly you can choose the second option.

The best approach is to take the nuts and/or bolts of each part and keep them in a labeled small plastic bag. After disassembly, you can then clean the fasteners and either keep them in the bag, or when possible, attach them to their respective part. On some items, you may want to clean them and tag them with

I like to work with an engine stand rather than on a bench. Note that this stand has a bar to leverage the engine around a 360° axis.

masking tape. Now, you'll have no trouble at all when it comes time to put them back on the block.

In my own work I do a combination of both. If the fastener has baked-on grease it goes into the hot tank. I want my new engine to look new; no greasy fingerprints from using grease covered bolts. If the interior bolts such as those associated with the head and rockers are clean, I wash them in solvent and proceed with the assembly.

Using an Engine Stand

In the previous chapter I just assumed the engine would be placed on a professional engine stand because that's what I use. You, however, may have decided to set the engine on a couple of four-by-fours while you decided if you wanted to work the engine on a bench or stand.

If you choose a bench it should be about 30-in. tall. Additionally, it should be built bomb-proof. Remember, that engine weighs some 600 to 700-lbs. fully equipped. The bench should also be large enough to turn the engine on its side. Some professional engine rebuilders prefer benches, but most, like myself, prefer an engine stand.

I like engine stands because you can work in a comfortable position and access all areas of the engine without having to wrestle with it. If you want to move the engine, simply rotate the stand—all allow for a 360° rotation, and most will allow you to set the engine in a variety of different angles.

If you didn't drain the oil earlier, you must do it now. Be sure to keep the used oil and coolant separate to meet EPA requirements for disposal.

Engine stands used to be quite expensive, but today they can be found at reasonable prices. The stand you see in the accompanying photos was purchased for just $50.00. Our previous stands, purchased in 1988, were $150.00. Since you'll likely be rebuilding your engine over a period of days, weeks or even months, it simply doesn't make "cents" to rent one. Believe me, the $50.00 will be money well spent.

REMOVING EXTERIOR HARDWARE

The first step is to remove all external parts, accessories, bracketry and the like, before you get into the engine itself.

Drain the Fluids

Even though you drained the coolant when you removed the radiator, there is still coolant in the waterjackets inside the block. To drain this, you must remove the two 9/16-in. bolts on each side of the block just aft of the engine mounts. These are extremely tight and will require the use of an impact wrench or breaker bar. For demonstration, I've used a breaker bar throughout.

When the bolt is removed, probably nothing will happen. The hole will more than likely be sealed with rust and scale. Use a screwdriver or punch to break through this scale, then watch out—coolant will come pouring out.

Over 99% of you will know how to drain the oil from the pan (for the remaining 1%, remove the plug from the oil pan) but many of you may not know what to do with the old oil. First of all, do not drain the oil into the same pan as the

To finish draining the water from the block you must remove the two engine block drain plugs (one on each side). You'll need a breaker bar for this operation.

This spark plug wire loom, retained by a head bolt, is in the way of removing the left engine mount. That's why I'm on it with a breaker bar.

While you're on this side, remove the dipstick tube. A gentle wiggle is all that's needed.

coolant. Both coolant and oil should be drained into separate containers and taken to your nearest EPA-approved disposal facility. Most gas stations with repair bays or one-stop lube and oil change places will dispose of your oil and coolant for a small fee, in some cases as little as $1.00. If you mix the coolant and oil, they won't be accepted for disposal and you've wasted your well-meaning effort. Please, don't dump the fluids into the ground or nearest trash bin. Our society can no longer afford such environmental carelessness.

Clutch & Flywheel

If your car is equipped with a manual transmission and you removed the engine with the bellhousing, this must be removed before you can bolt the engine to an engine stand. Remove the attaching bolts of the bellhousing and lift off. This will expose the clutch and flywheel.

To remove the clutch, loosen each mounting bolt a couple of turns at a time. This unloads the torque pressure evenly, preventing damage to the pressure plate cover. Once the pressure plate is loose, remove the bolts. Be careful! You're dealing with about 25 pounds of pressure plate and disc. WARNING: Do not get grease or even greasy fingerprints on the clutch friction surface. This will probably ruin the disc if it's reusable. Should you assemble the clutch with any grease spots you'll get clutch chatter, jumping and all kinds of unpleasant take-off problems. The only solution will be a new clutch plate.

After removing the pressure plate and disc, the flywheel is next. Remove the flywheel bolts. Be prepared to juggle about 40-lbs. of flywheel with sharp edges. I always use heavy leather gloves for this operation. If you don't want to use gloves, wrap a couple of shop towels around the edges to keep from cutting yourself.

Flexplate

If the engine has a flexplate instead of a flywheel, I like to leave it on until the engine is on the stand. It's a bear to remove while the engine is swinging

Because the mounts are retained with bolts and nuts, you'll need two wrenches or a wrench and socket to remove them.

The driver's side is finished with the removal of the spark plugs. I've left the oil pump until the engine is turned over.

I had no trouble removing the exhaust manifold. I did, though, use some penetrating oil. If you omit this step you run the risk of twisting off one of the studs. If you do, see the sidebar on page 94 to fix it.

around on the hoist. Once mounted to the stand, it can be removed with ease.

DISASSEMBLY PROCEDURES

The most efficient way to approach disassembly of the engine is to work one side at a time, then move to the front. Next, everything comes off the top. The engine is then turned over, the pan removed and all the moving parts are taken out from the bottom. That's the way we'll approach our project. For demonstration purposes, however, I'll sometimes jump back and forth.

Engine Mounts

The left (driver) side engine mount sits below a spark plug routing bracket which should be loosened or removed before removing the engine mounts. On the left-side engine mount, all three bolts are inserted from the outside. This puts all the nuts on the "inside" of the mount. On the right mount, however, it's just the opposite: all the bolts are inserted from the inside and the nuts are on the outside. This is cause for a bit of confusion. It can be avoided, however, by reinserting the bolts the way they came out. If you forget, I'll remind you of this during assembly.

If you look closely at the photos, or at your own project, you'll see the mounts are offset on each of their brackets. This means they are not interchangeable. Therefore, mark each one "right" and "left" as you remove it. If you get them reversed during assembly, which can easily happen, the engine will not sit on the frame properly and align with the bellhousing or tranny.

On the passenger's side, remove the fuel line and then the fuel pump. Note that I'm using my flare-nut wrenches.

On the passenger's side engine mount, the bolts all pass through from the inside. This is just the opposite of the driver's side.

If you didn't get the idler pulley before you pulled the engine, do it now to gain access to the water pump housing.

More spark plug wire looms to be removed. This one was retained by the front bolt of the exhaust manifold.

Exhaust Manifolds

Removing the exhaust manifolds is pretty straightforward. The only problem you might run into is stubborn nuts. High temperatures and moisture can cause rust problems. Rather than fight the problem or risk twisting out a stud, use a little penetrating oil, such as WD-40, on the threads. Should you break a stud you can find out how to repair the problem in Chapter 6 on *Reconditioning the Heads*.

Fuel Pump

Remove the fuel-pump-to-carburetor line. You can then remove the two bolts holding the fuel pump to the block.

FRONT END

Water Pump Housing

The first part of removing the water pump housing is to remove the water pump idler pulley. This is still secured with the top bolt. You removed the bottom bolt during the process of removing the belts.

Use a 9/16-in. socket on your ratchet to remove all but one of the bolts retaining the water pump housing. This is the bolt behind the heater-hose outlet. You'll need to use an open-end wrench on this bolt. There's not even room enough to get a box wrench in here.

You'll need to do a little prying and prodding with a screwdriver to break the seal between the housing and block.

All but the bolts behind the heater hose tube can be removed with your socket. You'll need a box-end wrench to get that one.

To get the water pump housing off, I pried between the housing and block with a screwdriver. Be careful not to scratch or mar the mating surfaces when prying with a sharp object.

A "little pressure" won't work with the damper bolt. Use your biggest breaker bar to overcome the 185 ft-lbs. of torque this bolt is holding.

Crankshaft Pulley & Harmonic Balancer

Begin by removing the six bolts retaining the crankshaft pulley to the harmonic balancer—sometimes called the vibration damper or just damper. The pulley may require some gentle persuasion with a screwdriver before it comes off.

To remove the harmonic balancer (damper) you'll need a 1-1/4-in. socket, a breaker bar at least 18-in. long and a harmonic balancer or steering wheel puller. Don't use a gear or pulley puller to remove the harmonic balancer.

The harmonic balancer is actually three parts—an inner wheel keyed to the crankshaft nose, a damping ring over the inner wheel and a rubber insulator bonded between the two. Using a standard pulley or gear puller will tear the outer ring off and destroy the damper assembly. Again, rent this tool if it's not part of your tool kit.

With the 1-1/4-in. socket and breaker bar, remove the bolt and washer holding the harmonic balancer to the crankshaft. A bar or large screwdriver bridging two bolts threaded into the crankshaft flywheel flange will keep the crank from turning. If the flexplate is still on, shove a block of wood or a screwdriver through one of the "ears" and brace it against the block or engine stand. Set the washer aside and reinstall the bolt. This will prevent damage to the damper-bolt threads in the end of the crankshaft. Mount the steering wheel puller according to the directions provided or as shown in the photo below.

Having removed all the bolts from the crankshaft pulley, it pops right off with a little screwdriver pressure.

I've seen more than one harmonic balancer destroyed by using the wrong puller or trying to knock it off with a hammer. Unless you have a puller that looks like the one above, buy or rent one to use on your project.

Here are the two pan bolts that are sometimes overlooked when removing the timing chain cover. Don't make this mistake.

It's too bad the gear is worn out. The chain does not exceed the 1/2-in. flex limit and could therefore be reused. I'll replace both parts, however. It will be worth it in the long run.

As the center-puller bolt is "tightened," the beam pulls against the two retainer bolts and pulls the damper off the crankshaft. Be careful not to lose the Woodruff key in the crank nose. It locks, or "keys," the harmonic balancer to the crankshaft. Under no circumstances should you try to beat or pry the damper off the crankshaft! The results will be a useless damper, or possibly a damaged crank.

Timing Chain Cover

With the harmonic balancer off, you have access to the timing chain cover. There's a trick here. There are two bolts holding this cover to the oil pan. These are often caked with grease and overlooked. You know what will happen if you try to pry the front cover off while it's still attached to the pan, so don't forget them. Remove the cover and whatever pieces of the gasket that come off with it.

Timing Chain & Sprockets

If a timing chain has more than 50,000 miles on it, it should be replaced. But if you are unsure, there is a test you can perform to check for wear and excessive slack before you remove the timing chain and its gears. Start by tightening one side of the chain by rotating the cam sprocket. Now, lay a straight-edge along the slack side of the chain. Can you deflect it

Remove the oil slinger to complete access to the timing chain.

The white spots you see on the cam sprocket are areas where the nylon covering over the aluminum gear has deteriorated away. This gear is now trash.

I was able to slide the gear and chain off with just a bit of pressure from behind, near the hub of the gear.

more than a 1/2-in.? If so, just throw the chain away. Now, check the cam sprocket for broken or excessively worn teeth. If there is wear here, it too becomes a candidate for the scrap pile.

Timing chains and cam sprockets have a relatively short life expectancy. They play a very important part in the performance of the engine. Because they are relatively inexpensive parts, they should be replaced if there is any indication of wear. A broken tooth will let the timing chain jump one or more teeth on the sprocket, putting the camshaft out of time with the crank. At best, cam timing will retard far enough to create a very poor running engine. At worst the valves will hit the pistons, bending the valves, notching the pistons and possibly destroying the heads. A very expensive lesson indeed.

The same warning holds for worn timing chains. Chain wear results in the chain becoming elongated. We sometimes erroneously call this "chain stretch." Therefore, the need for the deflection test. If there is more than 1/2-in. of deflection the chain has become elongated and may not be able to do its job. The additional length may cause the chain to climb or jump the sprocket teeth, retarding cam timing. Check these parts carefully. Finally, if the engine has an aluminum/nylon sprocket, remove it as a matter of course.

Removal of the cam sprocket is pretty straightforward. Having removed the oil slinger and the bolt that holds the sprocket to the cam, you should be able to slide the sprocket and chain off with a little gentle coaxing from a screwdriver. Apply the pressure as close to the camshaft (sprocket hub) as possible to avoid bending or cracking the sprocket. If this is not successful, try using two screwdrivers to apply equal pressure to both sides of the sprocket. For added leverage, place a nut between the blade of the screwdriver and the hub of the sprocket.

The intake manifold comes out with little effort.

With the sides and front clear, I turn my attention to the top. Begin by removing the valve covers. Usually a couple of smart raps with a soft (rubber) mallet does the job. These covers were sealed with silicone and required a bit of prying to break the seal.

DISASSEMBLING THE TOP SIDE

It's time to get inside the engine now and see what things look like. If the engine has been well maintained—regular oil changes, the use of a quality motor oil and no serious overheating—the insides will be clean and free of caked grease and carbon. The opposite is equally true. Poor maintenance will be reflected in a valley full of crud that may require a putty knife to remove before you can find the bolts! When I opened the project engine I found it to be quite clean.

The valley cover prevents trash from falling into the internal parts of the engine.

"Unload" the rocker arm assemblies a little at a time to prevent warping them from the pressure of compressed lifter springs.

Remove the rocker arm assembly, reinsert the bolts, secure them with a bit of wire, mark the unit "driver" or "passenger" and put it away.

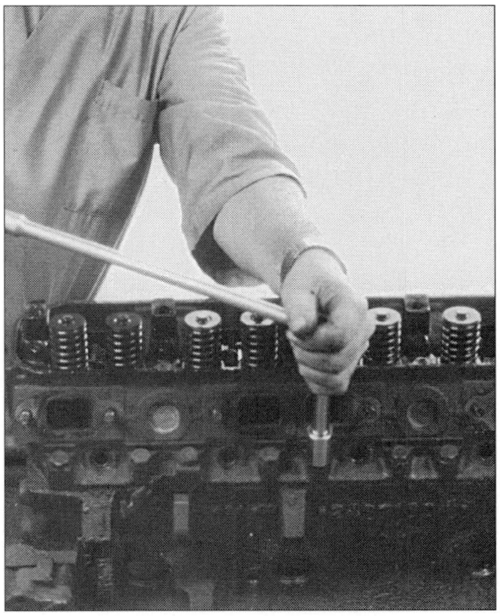

Back to the breaker bar. Unload the head bolts a little at a time as you did the rocker arms. This is the way they were tightened to full torque. It's simply good practice and prevents any possibility of warpage.

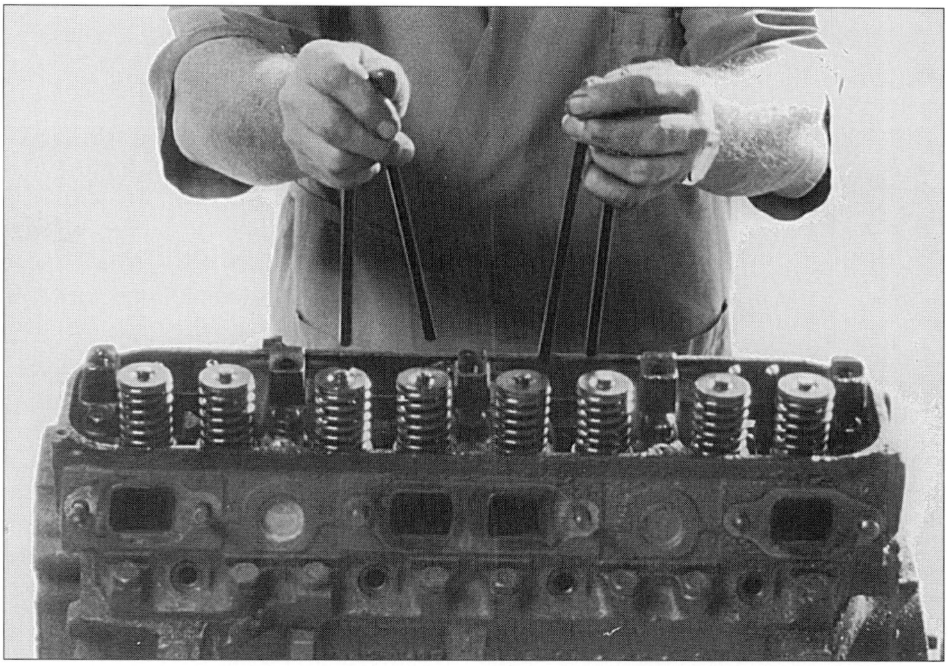

The pushrods may be mixed together. Discard any that are obviously bent. Later, you'll check for minor variance from true straightness.

Intake Manifold

Begin by removing both valve covers. A rap with a plastic mallet should loosen things enough to lift them off. If not, get out the trusty screwdriver again. Just be sure every bolt is removed from the valve cover. You can imagine what will happen if you pry up one corner of the valve cover while the other corner is still retained to the head.

Remove all of the manifold bolts. Here's one of those areas where it's going to be hard to replace the bolts after they've been removed. Each is a different length and all "bottom" into the heads. If you return a bolt that is too long for the hole it will "bottom out" before it reaches the correct torque. If the bolt is too short you run the risk of catching only two or three threads. When torqued, you might pull out these threads. If you have any worry about this, keep the bolts with the intake manifold.

After removing all the bolts, pry the intake manifold loose from the heads. Remove the valley cover and you'll see the lifters, pushrods and camshaft. Check the valley for foreign objects such as metal filings, chips and sludge.

Rocker Arms

The rocker arms must come off before anything else. This frees the pushrods and gives you access to the head bolts. The camshaft must be unloaded evenly and carefully to prevent bending or warping it. Remove the two outside bolts and the center one of the rocker arm shaft. Then remove the other two a couple of turns at a time. This prevents any undue pressure to either side from the loaded valve springs. Lift off the rocker arm assembly. Return the five bolts to their holes in the shaft and leave the rockers in place. Wrap some wire or tape around the bolts to keep them in place. Tag each assembly right or left and set them aside for future inspection. Remove the 16 pushrods. If any are obviously bent, throw them away. The rest may be checked later on. I use a 1-lb. coffee can to store them.

Heads

There are 17 bolts retaining the head to the block. Eight of these are on the bottom edge of the head, five are next to the valve springs and the remaining four are located along the back edge. Be sure all 17 bolts are removed before you

After returning one of the bolts to the head as a safety measure, use your pry bar under one of the bosses or through an intake or exhaust port to lift the head and break the seal. Remove the safety bolt and set the head aside. Although the heads are interchangeable from side to side, unless they will be completely rebuilt, mark them for location and return them to their original side.

attempt to pry the head from the block. Every mechanic knows someone who has tried to pry a head off without removing some of the bolts—usually those along the front edge. If you do this, the resulting two-piece head won't even make a suitable boat anchor! You also could cause irreparable damage to the block deck.

After removing all the head bolts, return one to the head and thread it in two or three turns. This will be your safety bolt when the head is pried loose. Insert a pry bar or a 10-in. extension into an intake port and pry up. Don't pry between the head and the block deck because this may scratch the deck or head surface. When the head breaks loose, remove the safety bolt and lift off the head. Be careful as the head weighs better than 50-lbs.

Although the heads look the same and appear to be interchangeable, you'll want them to go back to the side from which they came. Therefore, mark each one right and left then set them aside. Likewise, mark and save the gaskets. Later, we'll look at these with a critical eye for burns, blowouts, irregular torque and lots of other information. We'll discuss these areas fully in Chapter 6, *Reconditioning the Heads.*

Distributor Drive Gear

Before turning the engine over to work on the bottom end, remove the distributor drive gear. This gear is removed through the distributor-mounting hole. Insert a screwdriver blade in the distributor-shaft drive slot and rotate the gear counter-clockwise and lift it out.

Lifters

Although it is best to pull the lifters out when we remove the cam, there is the possibility that the lifters will fall out when you turn the engine over. It's always best to check first. Use a mechanic's magnet to try and pull each lifter out. If you don't succeed then you can be sure gravity won't pull it out either. If you get one out, be sure you mark it for location. Used lifters must go back on the same camshaft lobe. Refer to page 43 for a couple of methods to keep the lifters in order.

Using A Ridge Reamer

Before removing the pistons, it will be necessary to prep the cylinder bores. As each piston ring travels from BDC to TDC several million times during the life of the engine, it wears its bore. The space between the top compression ring at TDC and the block deck receives no wear. This unworn surface of the bore, and the carbon that collects there, is called the *ridge*. If you force the piston over the ridge, you may damage the piston ring lands and skirt. A broken ring is of little concern, but a damaged ring land makes a piston worthless.

So, if you plan to reuse the pistons the ridge must be removed. This is done with a tool called a *ridge reamer.* This adjustable tool works much like a wood plane. It fits and pilots in the top of the bore. Rotating the tool with a long lever and socket "planes" or "reams" the ridge off. Stop when the ridge is flush with the cylinder wall. Now the piston should slide freely out of its bore without damage.

This is a very clean engine. There is little carbon buildup and no sludge. The big problem is blocked water passages. This engine overheated and seized.

It is necessary to perform this operation on all eight bores. Thread the harmonic balancer bolt back into the crankshaft. Then rotate the crank with your socket and ratchet or breaker bar to move each piston down its bore so you can insert the ridge reamer. You will also have to rotate the crank when you remove the piston-and-rod assemblies, so leave the harmonic balancer bolt in place for the moment.

This is a fairly standard ridge reamer. The one blade you see on the left does the cutting while two adjustable guides keep the unit square in the bore. The little nut on the top adjusts the guide and cutter while the larger nut, with a wrench, is used to turn the tool.

To use the ridge reamer, adjust the top nut until the unit fits in the bore. Lubricate the blade with cutting oil or WD-40. With a box-end wrench, rotate the tool inside the bore so the cutting blade cuts into the ridge. As metal is cut away, continue to adjust the blade out. Stop just before you begin to cut the metal below the ridge.

DISASSEMBLING THE BOTTOM END

Now it's time to turn the block over and tackle the bottom end. If you're using an engine stand, all kinds of things will come falling, pouring and dribbling out when you turn the engine over. These include oil, grease, bits of gasket, maybe some more water and all manner of junk. So, place a lot of newspaper, towels, rags or an oil drip pan on the floor to catch this stuff. If you're working on a bench, do whatever is needed to protect it.

The bottom end of an engine has the greatest number of highly machined surfaces. These are extremely sharp and can quickly take off a knuckle or leave a nasty cut. Be forewarned and take precautions.

Oil Pump

We've left this exterior part on until this point, because it's easier to access one of the bolts hidden next to the block. Remove the bolts retaining the pump to the block. If you can rotate the pump, all the bolts are out. Now take a screwdriver or pry bar and gently remove the oil pump. It will slide out as soon as you break the surface tension of the oil between the block and pump.

With the engine upside-down you can easily access all the bolts to the oil pump. It pulls right out. It may need some coaxing to break the vacuum seal created by the oil on close tolerances of the machined surfaces.

Oil Pan

Jump right in and remove all the pan bolts. A speed wrench will move the job along quicker. Now you have a view of the oil pickup, connecting rods and crankshaft. This is the time to remove the oil pickup.

The pickup screws into the block. Under most conditions it can be removed by hand simply by unscrewing it. If, like the project engine, this doesn't happen, you'll have to resort to a pair of water-pump (channel-lock) pliers. Be very careful not to bend, crush or twist the pickup tube. The project pickup came out with just a little coaxing.

Check the pump pickup screen for foreign matter. You'll probably find bits and pieces of valve seals and maybe some timing chain sprocket chunks.

The pickup tube is screwed into the block and may need some encouragement from your channel-lock pliers to get it out.

Yes, it will come out. Take a look inside, next to the screen. You'll probably find a lot of mysterious particles.

Windage Tray

If your engine is a 413, 426 or 440 you'll find a windage tray under the crankshaft (over the crankshaft as you are now viewing it). The windage tray prevents the oil from being whipped up—like a chocolate shake. This translates into more power and less oil aeration. Remove the windage tray and set aside.

Rod Caps, Rods & Pistons

Begin removing the pistons from their bores by checking to see if they are correctly marked for location, or even marked at all. Rotate the crankshaft until the number 1 piston is at BDC. This puts its cap at a position where you can easily look for the number. Check the machined surface on the outside edge of the cap where it joins the rod. There should be a "1" stamped onto the surface of the cap and the corresponding surface of the rod. This number allows you to match cap and rod then return the assembled pair to their corresponding cylinders.

All of the operations involved in removing the pistons will require turning the crank. Install the 1-1/4-in. bolt you removed earlier. Now you can turn the crank with your breaker bar.

On the left side, left bank you should find numbers 1-3-5-7. On the right side, right bank the numbers 2-4-6-8. This allows you to return the rod and its cap to its correct crankshaft journal and facing in the right direction.

If the rods and caps are not numbered and get mixed up, their bearing bores must be reconditioned, or machined, to prevent unusual bearing wear or destruction. If you have a number-stamping set, mark each cap and rod missing its number as illustrated, before removing the rod cap. An electric engraving tool also works well. If neither of these are available, use a sharp punch to make a corresponding number of dots, i.e. two for number 2, three for number 3, etc.

Remove the nuts holding the cap to the rod. You probably won't be able to lift the cap off the rod. Don't try to pry it off as you did the heads and manifold.

If you find any of the rods and caps without numbers, this is one way to correct the problem. Use a center punch to mark the missing number: 2 dots for #2, 3 dots for #3, until all are correctly numbered.

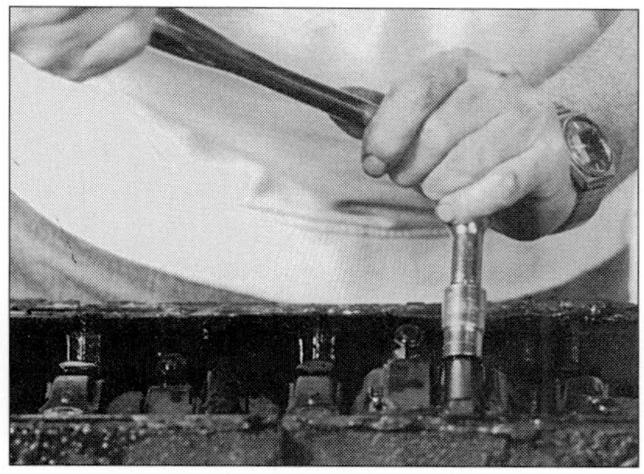

The easiest way to remove the rod cap nuts is with the rod at bottom dead center (BDC).

Grasp the cap with your thumb on one side of the balance pad—the machined boss at the bottom of the cap—and one or two fingers on the other side of the cap. Using a small hammer or plastic mallet, tap the cap gently on the side held by your thumb while attempting to rock it away from the side being rapped. Don't strike the cap with enough force to mar it.

Use a little care and forethought to protect the crankshaft journals and cylinder walls from nicks and scratches. Before pushing the piston out, cover each rod bolt with a short piece of hose or a sleeve made specifically for this purpose. These are often available as giveaways from parts stores if you ask.

Place a long hammer handle or piece of 2x2 wood against the piston's bottom side and push the piston out of the top of its bore. You'll need a free hand to catch the piston-and-rod assembly as it comes out. Unfortunately, you'll have to be a contortionist to do this. Short of having one arm that's 3-ft. long, I suggest you find a friend willing to help you here. Return the cap and nuts to their rods, matching the numbers on the machined surface. Don't remove the bearing inserts yet. They'll be used to determine the condition of the rods and crankshaft.

Repeat the process for all eight cylinders. Remember: make sure each cap and rod is numbered to its corresponding cylinder before loosening the nuts. Replace the caps on the rods

This photo is posed! When I actually pushed the piston out I had two helpers: one to hold the rod away from the sides of the cylinder bore and one to catch the piston as it came shooting out. With all this help we still dropped one on the floor and broke its skirt! Be real careful with this part of the operation.

This is a good time to remove the flexplate. I've shoved my 2x2 into position so the crank couldn't move. Now I loosened the bolts with my extra-long box-end wrench. These bolts are torqued to 55 ft-lbs.

The rear bearing seal retainer must be removed to access the number 5 main bearing cap.

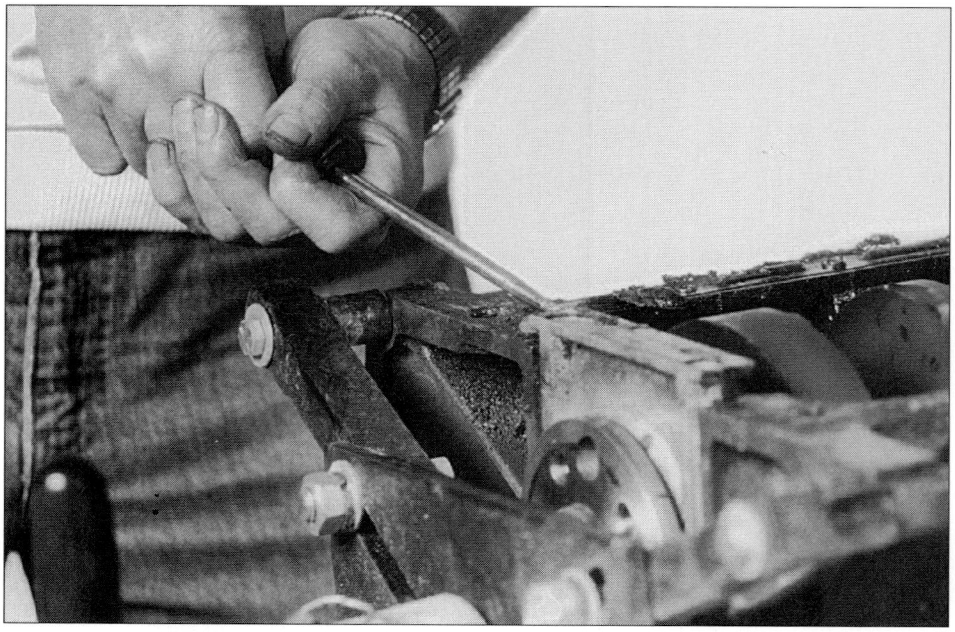

Like so many of the other parts, the rear seal must be "encouraged" with a screwdriver.

after each piston and rod is out. Keep the bearing inserts in place.

If you haven't done so, now is the last chance to easily remove the flexplate or flywheel. A block of wood between a crankshaft counter-balance and the block will keep the crank from turning so you can loosen the bolts.

Main Bearing Caps

Five main bearing caps retain the crankshaft.These are similar to, but larger than, the rod bearing caps. The main bearing caps are numbered front-to-rear, 1 through 5. If they are not, number them now in the same fashion as the rod caps.

The best way is to number the side of each cap and put the corresponding number on the same side of the block. If you overlook this, the block and caps will require align boring or honing—a relatively expensive machining operation. Remove the number 1 main cap bolts. You'll need your longest breaker bar because these bolts are torqued extremely tight—about 85 to 100 ft-lbs. Tap the main caps to loosen them just like you did the rod caps. Again, save the bearing inserts and keep them with their caps. They too will give you much needed information.

Before you can remove the number 5 cap, you must remove the crankshaft rear-bearing seal retainer (rear seal). You'll need a 7/16-in., 12-point socket for this operation. Remove the two bolts retaining this seal and carefully pry it out of the block, one side at a time. Work it up with the screwdriver then finish removing it by hand. Now you can get to the last main bearing cap.

The crankshaft weighs better than 50-lb. Be prepared for this amount of weight. Carefully lift the crank from the block, taking care not to bump any of the journals or parts of the block.

A crankshaft is one of the toughest, yet most delicate of all engine parts. Bearing-journal surfaces must be kept perfectly smooth. Nicks and scratches or rust can ruin it. It will withstand billions of blows from piston-and-rod forces, but drop it and you can damage the bearing journals or put it out of alignment. In fact, if you leave it lying on the garage floor or leaning against the wall for an extended period of time, it can warp.

To avoid this possible disaster, grease the bearing surfaces thoroughly, wrap it in an old blanket or trash bag and store it out of harm's way. Stand it on the flywheel or flexplate flange and wire it to a wall stud.

Remove the main bearings inserts and label them according to their position. Also indicate block half and cap half.

Main bearing caps are next. They too must be marked and returned to their exact same position. Any deviance and you will have to align bore the block and caps.

Be very, very careful lifting this heavy crankshaft out of the block. At about 50-lbs., it will certainly hurt if you drop it on any part of your body.

They will be used later during crankshaft inspection. Return the main bearing caps and bolts to their respective locations in the block with the bolts finger-tight. The main bearing caps go with the block when it is sent out to be reconditioned and machined. Again, it is extremely important to keep these caps and bearings in their original positions. If they get mixed up, the block and caps will require machining.

Camshaft

You're now ready to turn your attention to the cam. However, you must first remove the fuel pump actuator. This rides on the front-most lobe of the cam and actuates the fuel pump. With the block in the same position you've had it, you'll find what looks like an oil gallery plug beneath the machined mounting surface for the fuel pump. This is not a gallery plug but an access hole sealed with a hex-head bolt. This prevents the actuator from falling out of the block when it's in the right-side-up position and the fuel pump is removed. Using a hex driver, remove the bolt and slide the fuel pump actuator out of its bore.

To extract the camshaft and avoid damaging the lifters, be sure each lifter is in the up position as far as it will go. This is easily accomplished by rotating the cam a few turns using the camshaft sprocket locating dowel as a crank handle.

I have found the easiest way to remove a camshaft is to begin by screwing a 4-in. bolt into the nose. This acts as a handle and gives you something to hold on to. Remove the camshaft slowly and gently. Pull with one hand and guide with the other by reaching down between the main bearing webs. The cam and lifters are the most delicate engine components. If the camshaft and lifters are reusable, you'll want to avoid damaging them.

Camshaft bearing material is similar to rod and crankshaft bearing material. It damages easily, which then makes the bearings unusable. Even so, cam bearings receive very little wear and can be reused if not damaged. So guide the cam out carefully!

The camshaft is similar to the crankshaft in that it will suffer the same fate if abused. Surface damage will make it unusable, and it will also warp.

Before removing the fuel pump actuator, you must remove the retainer bolt, below the fuel pump deck surface. This bolt is there to prevent the actuator from falling out when the fuel pump is removed.

Therefore, wrap it up for now and hang it or stand it on end. In the inspection and reconditioning chapter we'll determine if the cam is reusable or if a new one should be purchased.

Lifters

The last mechanical engine parts to be removed are the valve lifters. If you intend to reuse them, then like all the other reusable parts, take care removing them. Keep the lifters in order as they are removed. I used to use egg cartons for this. One was marked right and the other left. Then, each compartment was numbered 1 through 8. This worked very well until one day I knocked one of my cartons of "eggs" off onto the floor. Eight valve lifters trashed because I didn't know which was which. Now, I wrap them individually in wide masking tape and write their location on the outside.

Varnish buildup at the camshaft end of the lifters makes it difficult (or impossible) to slide them out the top of their bores, so don't even try. Push them out the bottom, through the space formerly occupied by the cam. If this doesn't work, use a section of wood dowel that fits over the outside edge of the lifter and use this to push it out. If you have to use a hammer to tap them out, be sure you stay away from the very

You'll need a hex driver like one of these to access the fuel pump actuator and to remove the oil gallery plugs.

center of the lifter where the pushrod locates. This can damage the plunger, retainer spring or pushrod seat.

After all the lifters are out and safely put away, turn your attention to removing the core plugs (freeze plugs) and oil gallery plugs.

Core Plugs

There are six core plugs to be removed, three to a side. They should be removed to allow cleaning solution to flow through the waterjackets, removing the accumulated deposits of rust and corrosion. If you use the following procedure, core plug removal is a snap.

With a large punch, tap smartly on one side of the plug. This will rotate it 90° in its hole. Using a pair of channel-lock pliers, grab the outside edge of the plug. Use the heel of the pliers as a fulcrum and pry the plug out. Simple as can be! If you should accidentally push a plug all the way into the block, don't worry. Simply reach in, pull it out as far as possible, then use the pliers to finish the job.

Oil Gallery Plugs

There are four oil gallery plugs to deal with: two in the front, behind the timing chain and gears and two in the rear, behind the flexplate. Get out your hex drivers and remove them. This gives you access to clean the oil galleries as will be demonstrated in Chapter 5 on *Reconditioning the Short Block.* ■

The arrow is pointing to the camshaft sprocket locating dowel. Use this as a handle to rotate the cam, causing all of the lifters to raise in their bores. The 4-in. bolt acts as a handle when you remove the cam.

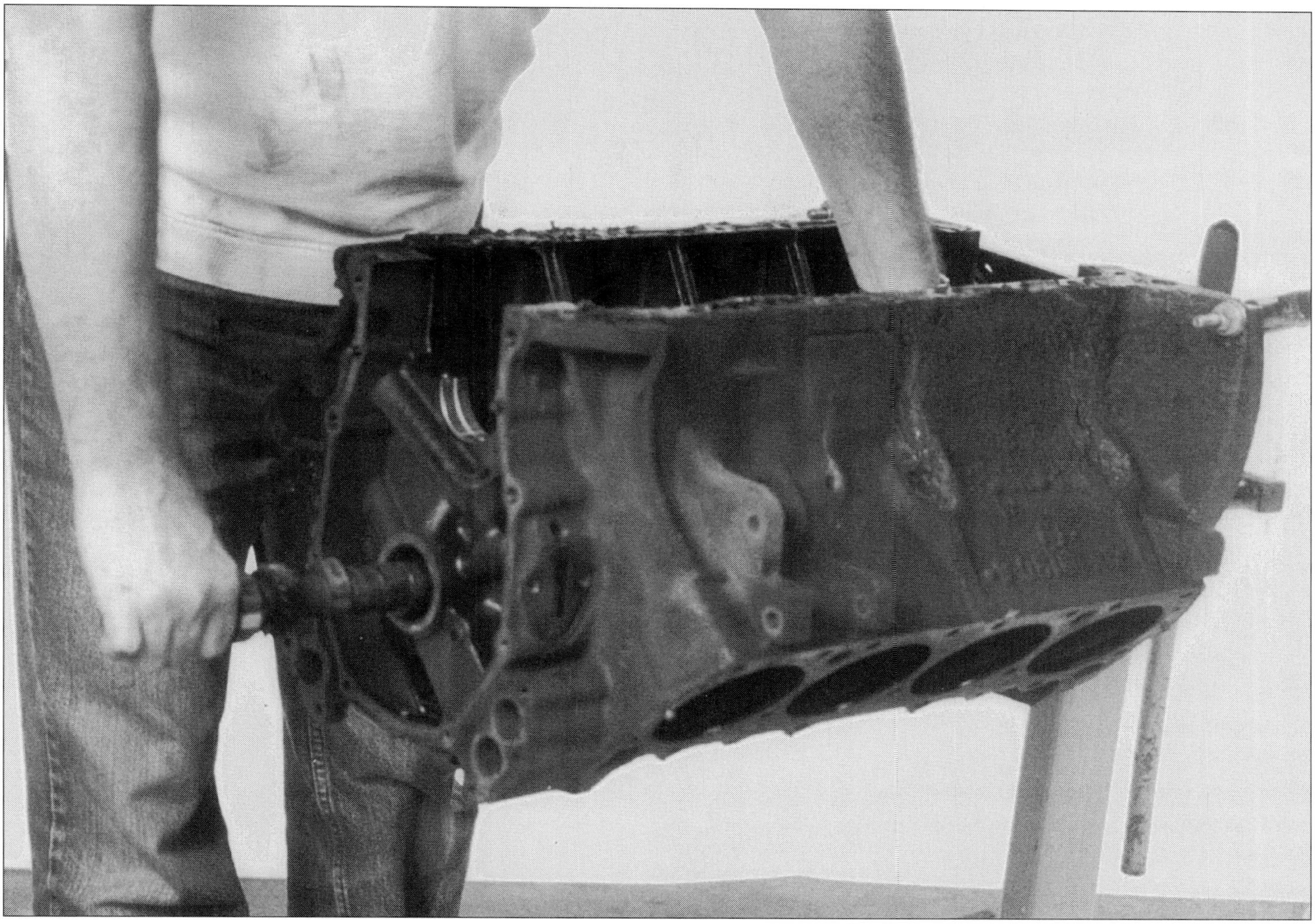

With my right hand on the bolt in the front of the cam and my left inside the block, I guide the cam out—eliminating the possibility of scratching any of the cam bearings.

The best way I know of to remove the core plugs. Use a punch to rotate them in their bores. Then leverage them out with vice grips or your channel-locks.

PARTS IDENTIFICATION & INTERCHANGE

4

The big-block Mopar engine builder is in an enviable position when it comes to finding parts; almost all of them are interchangeable. It would probably make this chapter shorter if I explained what couldn't be interchanged. I intend, however, to add a bit of history for you as we go along. This will make the chapter longer and you wiser!

It is in the finest American tradition that the engine builder borrow parts from one engine to make another engine operate, work better or go faster. Not only do you and I borrow, but so does the manufacturer. This is the case for the "B/RB" engines. Many of the parts are interchangeable from one engine to another. There is a danger inherent in this, however. Sometimes a part will fit but it won't work properly.

In addition to helping you find the right part to fit your engine, this chapter will help you learn to identify what a part is. If you go wading through a field of junked blocks how do you know what started life as a 383 or 400? Is that crankshaft in the back of the junkyard the one you're looking for? Did you just throw away a stock piston or was it an aftermarket? Is that the kind of thing you'd like to know? Read on, then. These and many other mysteries will be unraveled!

IDENTIFICATION NUMBERS

There are three basic pieces of information you need to identify your engine: displacement, year of manufacture and engine number. The engine number is all the information that's needed for positive engine identification.

Assuming the engine you removed from the car was original equipment, this information can be found on the Vehicle Identification Plate (VIP).

Until 1967, the VIP was located on the driver's door post. In 1968 it was moved to the left top of the instrument panel, at the base of and visible through the windshield. This tag is stamped with 13 digits that give car line, price class, body type, engine displacement, model year, assembly plant and sequence number—the car's position on the assembly line the day it was built. This information becomes the car's serial number. The chart on page 48 gives all the information needed to "read" your Vehicle Identification Number.

Engine Block Stamping Number

Another number you'll need is the engine block stamping. This number is stamped into the engine block and gives the code for engine displacement, year and month of production, performance level and special equipment. The "B" block or 383 and 400 cid, stamping is found on the right side of the block adjacent to the distributor. The 413, 426 and 440 cid, or "RB" series can be identified by the stamping on a pad at the left front of the engine in front of the valley cover. The chart on page 49 tells you how to interpret this number.

Casting Number

When a part is cast, be it a head, block or crank, the casting number is incorporated into the casting. It can help identify a part, but it is not necessarily adequate for complete identification. For complete identification you'll also need the engine size, year of production and stamping number.

Particularly important are the casting numbers for the heads. These numbers distinguish between years of production and, consequently, different configurations.

Armed with engine size, year, casting and engine stamping numbers, you can identify various engine components. This information, and a physical description, will help you find out if one part is interchangeable with another. The various tables nearby will help you decipher these numbers.

VEHICLE ID PLATE INTERPRETATION*

Car line

B	Barracuda
C	Chrysler
	Newport/New Yorker, 77—80
D	Dodge, 73—74 Monaco
E	St. Regis
F	Lebaron
G	Diplomat
H	Volare
J	Challenger
L	Dart
M	Horizon
N	Aspen
P	Plymouth Fury, 77—80 Grand Fury
R	Satellite, 75—80 Plymouth Fury
S	Cordoba
T	Newport/New Yorker
V	Valiant
W	Coronet/Charger, 70—80 Monaco
X	Charger SE—Magnum SE
Y	Imperial
Z	Omni

Price Class

E	Economy
G	Dodge Taxi
G	Grand, 72 and later
H	High
K	Police
L	Low
M	Medium
N	New York Taxi
O	Super stock
P	Premium
S	Special
T	Taxi
X	Fastback

Body Type

21	2-door sedan
22	2-door special coupe
23	2-door hardtop
27	Convertible
29	2-door sports hardtop
41	4-door sedan
43	4-door hardtop
45	6-passenger wagon
46	9-passenger wagon

Sequence Number

1 00001
per day, starting with 1 00001,
1 00002, -3 and so on

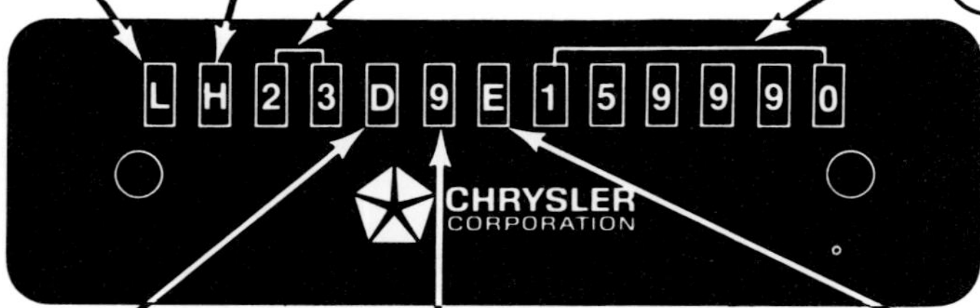

Engine Displacement

A	170 CID (60—69)
	1.7 liter (78—80)
B	225 (64—69)
	198 CID (70—80)
C	Special order 6 (60—69)
	225 (70—80)
D	273-2Bbl* (64—67)
	225-2Bbl (77—80)
E	273-4Bbl (64—67)
	Special order 6 (70—80)
F	318 Polysphere (60—66)
G	383 (64—66)
	318-2Bbl (67—80)
H	383 Hiperf (64—67)
	340-4Bbl (68—73)
	318-4Bbl (78—80)
J	426 Hemi (64—68)
	340-6Bbl (71)
	360-4Bbl (73—80)
K	440 (64—67)
	360 (70—80)
L	440 Hiperf (64—67)
	383-2Bbl (67—71)
	360-4Bbl Hiperf (74 & 78—80)
M	400 (72—77)
N	383-4Bbl (68—71)
	400-4Bbl (73—80)
P	400-4Bbl Hiperf (68—78)
R	426 Hemi (68—71)
	360 (78—80)
T	440 (68—78)
U	440-4Bbl Hiperf (68—78)
V	440-6Bbl (69—72)
Z	Special order 8

*Bbl is abbreviation for carburetor barrel.

Model year code

6	1966
7	1967
8	1968
9	1969
0	1970
1	1971
2	1972
3	1973
4	1974
5	1975
6	1976
7	1977
8	1978
9	1979
0	1980
1	1981

Assembly Plant

1 or A	Lynch Road
2 or B	Hamtramck
3 or C	Jefferson
4 or D	Belvidere
5 or E	Los Angeles
6 or F	Newark
7 or G	St. Louis
8	Export
H	New Stanton
P	Wyoming (Export)
9 or R	Windsor

*Pre-'66 models used a ten digit VIN that did not include engine displacement or price class.

BLOCK SERIAL NUMBER STAMPINGS

There are two locations to look for block serial numbers on the B and RB engines: on a pad at the left front of the engine on top as noted in the body copy on page 47, and also on the very bottom of the engine next to the oil pan flange. The stampings on the top of the block will look something like this:

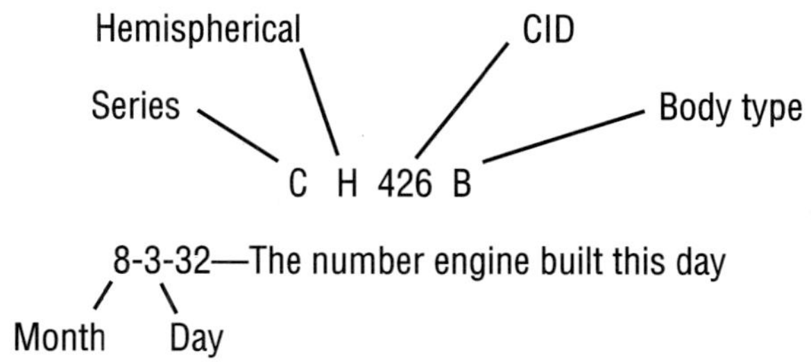

You may also find at this location other stampings such as a Maltese Cross, WT, or an X. These markings are noted below. The stamping on the bottom of the block will be the formal serial number for that block and read similar to the following: 9-10-68 PT38326010001. These numbers are decoded as follows:

A. The first numbers are the date of stamping (9-10-68).
B. The next letters indicate the engine assembly plant—in the case of our example, Trenton.
C. The next three digits indicate cubic inch displacement: 383 cid.
D. The next four digits (2601) are the 10,000 day calender numbers: 2601 = Sept 10, 1968.
E. The final four numbers (0001) show that this was the first engine assembled that day.

SERIES:

S = 1962	A = 1965	D = 1968	G = 1971
T = 1963	B = 1966	E = 1969	H = 1972
V = 1964	C = 1967	F = 1970	J = 1973

OTHER STAMPINGS:

Diamond = 0.008-in. oversize tappets
Maltese Cross = 0.001-in. undersized crankshaft
Maltese Cross and an X = 0.010 undersized crankshaft
A = 0.020-in. oversized cylinder bore
B = 0.010-in. undersized main and rod bearings
E = cast crankshaft
H = standard 4bbl
HP = high performance
LC = low compression
O.S. = 0.005 oversized valve stems
P = premium fuel recommended
R = regular grade fuel may be used
S = special engine
WT or TW = water test
X = oversized valve guides

ASSEMBLY PLANTS

K = Toluca	M = Mound Road	MV or MN = Marysville
PT = Trenton	T = Trenton (400 and 440 only)	

BLOCK CASTING NUMBERS				
CASTING #	CID	YEAR	TYPE	REMARKS
1852029	413	1962	RB	Max Performance & Truck
2120529	413	1959-65	RB	
2205697	413	1959-65	RB	Passenger Car
2205697	426	1964	RB	Passenger Car
2406730	426	1963-65	RB	Wedge & Max Performance
2468030	413		RB	
2568130	383	1959-71	B	
2468330	426	1964-71	RB	Hemi Head
2432230	426	1964-66	RB	Wedge & Max Performance
2536430	440	1966-72	RB	
2658836	426		RB	Industrial
3614230	400	1971-72	B	
3698630	400	1973-78	RB	
3698830-440	440	1973-78	RB	
4006530-400	400	1976-78	B	
4006630-440	440	1978	RB	

ENGINE BLOCK IDENTIFICATION

The term cubic inch displacement (cid) will be used quite frequently throughout this book. Therefore, a clear understanding of what cid means will be useful. An engine's size is determined by its displacement, the volume displaced by all of the pistons of an engine as they travel once from the top of their bores to the bottom. For example, on a 400 cid engine it would take 400 cubic inches of liquid to fill all eight cylinders from the top of each piston at the bottom of its travel—bottom dead center or BDC—so the liquid would be level with the top of the piston at the top of its travel—top dead center or TDC. The liquid in our 400 example would be 4.34-in. wide by 3.38-in. deep (bore X stroke). Let's use these two figures in a formula to discover how 400 cubic inches was determined. The formula for displacement is:

Displacement = pi/4 x bore2 x stroke x number of cylinders

The mathematical constant *pi* equals 3.1415927, and divided by 4 the value is 0.7854. If we plug in the above numbers of our 400 cid engine, we get:

Displacement = 0.7854 X 18.835 X 3.38 X 8
Displacement = 400 cubic inches.

The Chrysler big-block family consists of five basic block sizes identified by their cid: 383, 400, 413, 426 and 440. This grouping is called the "B" series and is further refined into the "B" and "RB" series. Refer to the chart below to recall what these designations mean.

As a family, the "B" series has the following in common: They all weigh about 225 pounds and they each have a common cylinder-to-cylinder bore spacing of 4.80-in. As stock, the low blocks share a stroke of 3.38-in. and the raised block incorporates a 3.75-in. stroke. Let's look at them a bit more closely.

"B" Series 383 and 400 cid

If you go back a zillion years, actually to 1956, Chrysler introduced its "A" series of engines—now referred to as "LA" engines—and composed the series we call the small-block. In 1958 Chrysler introduced its big-block engines, the 350 and 361. If they already had an "A" engine what would be the logical nomenclature for the next? You got it, the "B" engine! I'm not sure if this is true or not, but it certainly sounds logical to me.

The original "B" series engines were the 361, 383 and 400. The 361 sort of slipped into oblivion as a big-block but that displacement appeared as a small-block 360 in 1971. For our discussion we'll only consider the 383 and 400 cid engines.

These are the same block castings but with bores cored differently for the two different sizes. The 383 has a bore of 4.25-in. and the 400, 4.34-in. Both have a stroke of 3.38-in.

	Bore		Stroke	CID
Lo Block	4.12"	x	3.38"	361
	4.25"	x	3.38"	383
	4.34"	x	3.38"	400
Raised Block	4.18"	x	3.75"	413
	4.25"	x	3.75"	426
	4.32"	x	3.75"	440

This chart will help you understand the difference between the "B" (Lo Block) and "RB" (Raised Block) series of big-block Mopar engines.

CONDENSED HISTORY

383:	1-2bbl (RB) was offered from 1959-1960		
383:	1-2bbl (B) was offered from 1961-1971		
383:	1-4bbl was offered from 1959-1971		
383:	2-4bbl was a dealer installed option		
400:	1-2bbl was offered from 1971-1978		
400:	1-4bbl was offered from 1972-1978		
413:	1-4bbl was offered from 1959-1965		
413:	2-4bbl was offered from 1959-1965		
426:	1-4bbl was offered from 1963-1966		
426:	2-4bbl was offered from 1963-1971		
440:	1-4bbl was offered from 1966-1978		
440:	3-2bbl was offered from 1969-1972		
413:	2-4bbl Max Wedge	Stage I	1962
426:	2-4bbl Max Wedge	Stage I	1963 Early
426:	2-4bbl Max Wedge	Stage II	1963 Late
426:	2-4bbl Max Wedge	Stage III	1964
426:	2-4bbl Race Hemi	A-864	1964
426:	2-4bbl Race Hemi	A-990	1965
426:	2-4bbl Street Hemi	Stage I	1966-1967
426:	2-4bbl Street Hemi	Stage II	1968-1969
426:	2-4bbl Street Hemi	Stage III	1970-1971

Chrysler indicates that both of these blocks may be overbored to 0.040-in. Mopar Performance, Chrysler's high performance parts division, indicates that these blocks may be bored to 0.060-in. without danger of making the cylinder walls too thin. However, if the engine is to be used for racing purposes the cylinder walls should be checked very carefully.

The 1976-'78 400 cid (and 440 cid) was a thin-wall casting and as such should be bored no greater than 0.020-in. over.

Mounts—The engine mounting bosses are the same on all "B/RB" engines. Therefore, the engines are interchangeable within a car designed to accommodate either one. In other words, you could install a 400 into a car from which you removed a 383. Because all the mounting bosses are the same, the above holds true for putting a 440 where a 413 came out. Now, here comes the "but."

This is true for all but the 426 Hemi block which has different engine mounting lugs and will therefore not interchange with a "B" engine wedge block.

"RB" Series 413, 426 & 440 cid

"RB" stands for raised block and refers to a modification of the block casting to allow for additional stroke length. The stroke was increased from 3.38 to 3.75-in. by increasing the cylinder bore length almost 3/4-in.—actually 0.745-in. This also increased the crankshaft centerline to the top of the block dimension from 9.98-in. to 10.725-in.

Chrysler wanted a strong block. To that end there are extensive support webs in the mid- and low-portion of the case. Sidewalls supporting the main bore extend below the crank centerline. Additional truss material joins the main webs to these sidewalls. By doing this it extends the block casting more than 2-in. below the crank centerline, allowing for a completely flat pan-rail surface.

It is important to say something here about the term "wedge." Wedge refers to the shape of the combustion chamber within the head of the engines we've been discussing. This helps us differentiate from that more famous combustion chamber, the hemispherical or "Hemi."

413 cid Max Wedge—As noted above, the 413 was the first of the "RB" series. The bore was increased to 4.18-in. and the longer stroke of 3.75-in. was achieved by extending the height of the bore. It was introduced in 1960 in the Chrysler 300E. This was a dual four-barrel, long-ram inducted, 10:1 compression engine, factory rated at 380 horsepower (at 5000 rpm) with 450 lbs-ft. of torque.

Two years later, 1962, marked the introduction of the 413 Max Wedge. This was a dual four-barrel, cross-ram engine with 420 horsepower. The 1963-1964 Max Wedge block had valve clearance notches machined in the block at the top of the cylinder bores. In that same year the 426 Wedge was introduced.

426 cid Wedge—By increasing the bore of the 413 Max Wedge from 4.18-in. to 4.25-in., 413 became the 426. Further modifications were introduced to make this a total performance package. A new casting was made incorporating larger oil feeds and notched bores to accommodate the 1 7/8-in. exhaust valves. Additionally, this engine incorporated further refinements of special rods, crankshaft, pistons, heads, valves, valve gear, intake manifolds, carburetors and exhaust manifolds.

440 cid—The biggest of the "brute force" blocks was introduced as the 440 in 1966. To do this the new block was

Forged or billet crankshafts may be identified by the wide flash mark running the full length of each throw *(arrow)*.

bored to 4.32-in. and retained the "RB" stroke of 3.75-in. In 1969 Chrysler added three, 2-barrel carburetors, special rods, crankshaft, timing chain, camshaft, valve springs and intake system. This engine has been affectionately termed the "440 Six-Pack."

Like the 400, the 440 was a thin-wall casting design and the same prohibition of boring the engine no more than 0.020-in. over holds here.

426 cid Hemi—I've separated the 426 Hemi from the wedges because it is quite different, yet still an "RB" block. Most of the Hemi parts are not interchangeable with a wedge motor. Even the engine mounts are in a different position, which means a Hemi motor can't be installed directly into an engine compartment originally equipped with a wedge engine. This is not to say the Hemi can't be installed in these cars—or others. It's done all the time. It just won't "drop in" like the others.

Here are a few of its stats. Introduced in 1964, the block differed radically from its stable mates. One of the most significant differences is an extension of the inner edge of the deck to provide support for four retaining studs on the top edge of each cylinder head. These four studs are secured with nuts tightened from inside the valve lifter (tappet) chamber. The block also differs in the area of oil drain-back holes from the rocker chamber.

Cross-bolted main bearing caps for numbers 2, 3 and 4 plus increased main bearing webs make this block different and "beefier." Main bearing inserts are grooved top and bottom for improved oil flow to the rods during high rpm. The Hemi block has a bore spacing of 4.80-in. Centerline of the crankshaft to the top of the block is 10.725-in. Like the other "B" and "RB" blocks the centerlines of the crank and cam are spaced at 5.15-in.

Like its brothers, the 426 Hemi can be overbored 0.060-in., provided it is checked carefully for core shifting.

426 Hemis destined for racing were stress-relieved in production. The stress-relieving process consisted of heating the block and main caps to 1050° F. and holding it at this temperature for 2 hours. It was then furnace-cooled to 500° at a rate of less than 200° per hour. Finally, it was air-cooled to room temperature. Stress-relieving insured straighter bores with less movement during use.

CRANKSHAFTS

Before entering into a discussion of the individual crankshafts, we should discuss the various types of crankshafts and how they differ from one another.

Essentially, there are three types of crankshafts: forged, cast and cast/shot-peened. Each type has specific load capabilities.

Cast cranks have a narrow parting line *(arrow)* rather than the flash mark.

The forged crank, or as it is sometimes called, a billet crank, begins life as just that: a straight bar, or billet of steel. This is heated to about 2000° F., then forged into the crank shape. It is then machined to the desired specs.

The second method of crank manufacture is casting. Here, a sand mold is made in the basic crank shape and molten iron is poured in. This is then machined to crank shape and specifications.

To create the third style, added strength is given to the casting by shot peening. In this process, the casting is submitted to a treatment much like sand blasting or glass beading, only steel shot is used with considerably more force.

The forged crank is the strongest of the three. The forged steel grain flows with the shape of the crank. The cast crank is the least strong of the three. It doesn't have a grain structure that flows with the shape of the crank as with the forged steel crankshaft. Rather, its crystalline structure is random. But shot peening compresses the surface, prestressing the casting so it is more resistant to cracking.

Crankshaft Interchange

The "B/RB" series of engines provides a veritable potpourri of displacement combinations because most of the cranks are completely interchangeable. At least 18 different engine displacements can be created through the use of various stroke lengths and block sizes. With only stock parts, displacement variations of more than 100 cubic inches may be obtained. Stroke lengths may be varied from 3.218-in. to 3.750-in. If you're into the performance end of engine building, your options are virtually unlimited.

However, the "B" and "RB" engines have different main bearing journals and are therefore not interchangeable. The "B" engine has the smaller main bearing journal at 2.625-in. as compared to the "RB" at 2.750-in. Both cranks incorporate the same rod journal diameter of 2.38-in. Now let's examine some features of each crankshaft type.

383 cid—The 383 engine started life with a forged crank, but didn't end that way. Late in 1971, the 383 2-barrel was given a cast crank. Cast crankshafts use a specially balanced vibration damper and a specially balanced torque converter or flywheel. This is called external balancing. All cast crankshafts were externally balanced. Refer to the drawing below to see how this is identified.

400 cid—Cast cranks were used in all 400 cid engines except some 1972-74 4-barrels. These were also cast but were of a much stronger casting.

413 & 426 cid Max Wedge—The Max Wedge crank was a forged steel, flame-hardened unit and was the "weapon of choice" in all high performance buildups.

440 cid—The widest variations in crankshafts are in the 440 series. The 440 4-barrel (1967-1973) used a forged crank of which there are several versions. The first, 1967-70, was a standard forged crank. A stronger unit was developed for

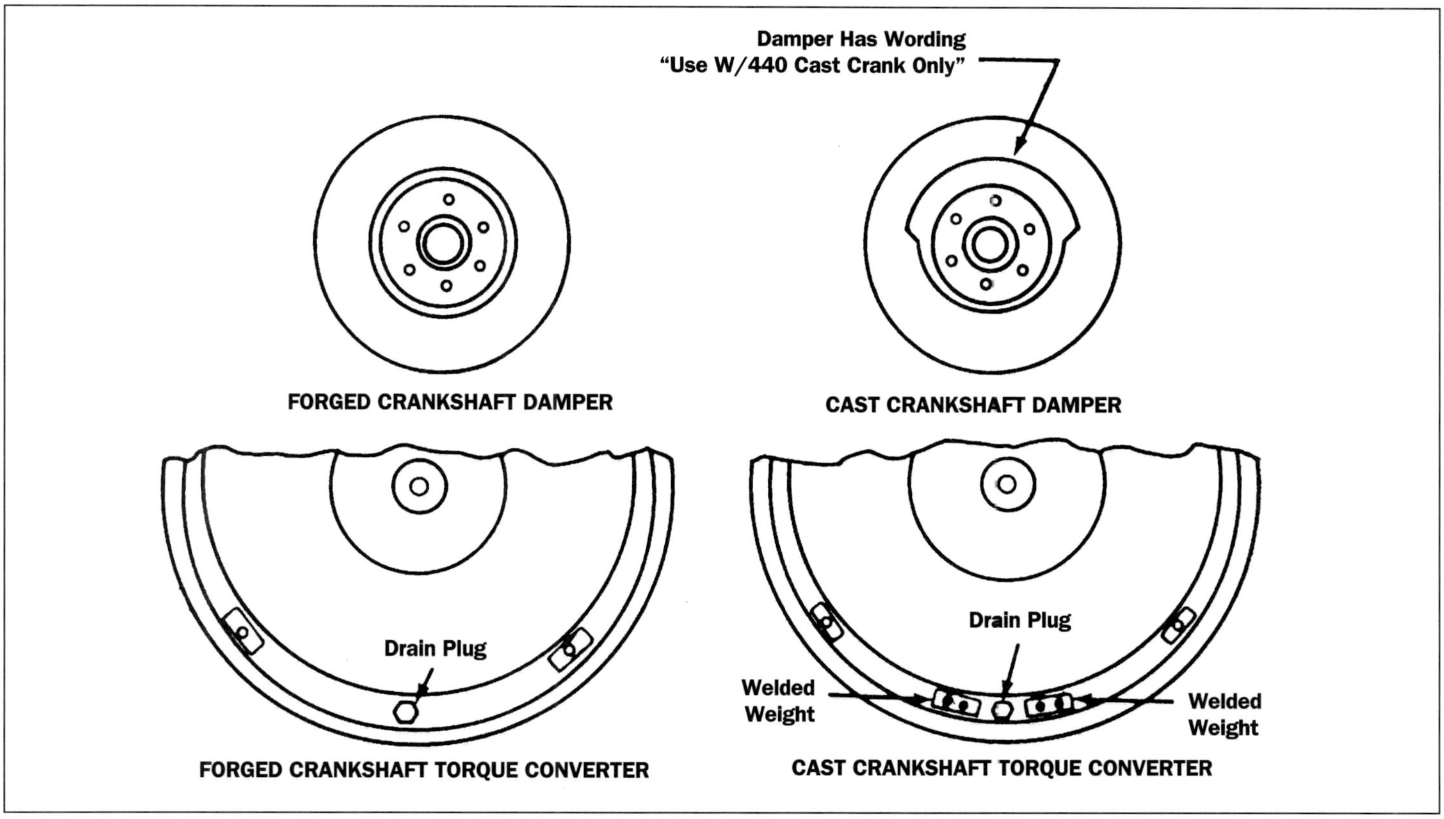

You can determine whether or not you have a cast or forged crank by checking the above diagram. Note the cast crank is externally balanced with weights on the torque converter

CRANKSHAFT VIBRATION DAMPER

P/Cast #	Years	CID	F/C	Remarks
2275897	1962-65	383 413	F	
2275897	1964-66	426	F	Wedge
2532105	1964	426	F	Hemi
2532795	1965	426	F	Hemi
2658457	1966	383 440	F	
2658457	1967-69	383 440	F	
2658457	1970-71	383-4bbl.	F	
2658457	1970-71	440	F	W/Low Performance
2658871	1966-71	426	F	Hemi NC#
3512017	1970-71	440	F	W/Hi Performance
3577180	1970-71	383-2bbl.	C	W/Auto Transmission
3577785	1973-74	440	C	W/Auto Trans & Low Perf
3614371	1972	400 440	F	
3614371	1972-74	400-4bbl.	F	W/Man Trans & Hi Perf
3614371	1974	400-2bbl.	F	
3614371	1973-74	440	F	W/Auto Trans
3614372	1972	440	F	W/Hi Performance
3614374	1972	400-2bbl.	C	W/Auto Trans
3614374	1973	400	C	W/Auto Trans
3830183	1975	440	C	W/Hi Performance
3830184	1975	400	C	
3830482	1975	440	C	W/Lo Performance

F = Forged
C = Cast
NC# = No casting number found on part

You can determine whether or not you have a cast or forged crank by checking the casting numbers on your damper (located on the front or rear) and referencing it to the above chart. Note the cast crank is externally balanced with weights on the torque converter. See drawing on previous page.

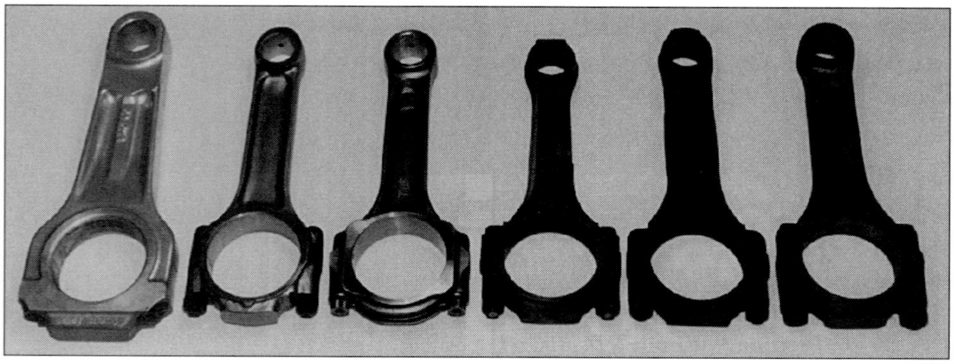

There's a wide variety of connecting rods for the engine builder to play with. *L to R:* Brooks aluminum rod, a "race-only" rod weighing 28 oz.; Six-Pack rod prepared by Keith Black—lightened, polished and shot-peened, 31 oz.; Childs and Albert premium forged steel, 32 oz.; stock 383-400, 26 oz.; stock Six-Pack at 33 oz.; stock street Hemi, 35 oz. *Photo courtesy* Chrysler Power *magazine.*

the 1972-73, 4-barrel high performance models, although there was also a standard crank in '72-'73 that was balanced for lighter pistons.

The 1970-'72 440 Six-Pack used a forged steel crank that utilized external balancing to compensate for the heavier pistons and rods. This was not a modified earlier forging but a brand-new unit made of premium quality steel.

The cast crank was introduced in 1974. This crank also requires external balancing .

426 Hemi—The Hemi crank may be interchanged with other "RB" cranks. However, Mopar Performance suggests that the appropriate Hemi connecting rod be used in conjunction with it. The Hemi crank also sports an 8-bolt flange while the "RB" cranks have a 6-bolt.

Final Crankshaft Thoughts

There is another crankshaft interchangeability consideration. Will the engine be bolted to an automatic or manual transmission? Cranks used with a manual transmission must have a pilot-bushing bore—a feature absent from the automatic.

CONNECTING RODS

Connecting rods for the "B" series engines are extremely strong, steel forgings of I-beam design, incorporating 3/8-in. cap bolts. Center-to-center lengths of the "B" and "RB" rods are 6.358-in. and 6.768-in. respectively.

Both "B" and "RB" engines use a pressed piston pin in the rod. Only the 426 Hemi used a full-floating wrist pin.

In 1970 and '71, the 440 Six-Pack engine came equipped with another special forging that had a wider beam section from the pin to the crank end. This high performance rod was also used in the 1971 to '73 440 four-barrel engine. A special high performance rod for the Max Wedge engine was developed but fell from favor because aftermarket

CONNECTING ROD PART AND CASTING NUMBER

Part #	Casting#	Year	CID	Remarks
1737690	1737692	1962	383	
1851533	1851535	1962-63	413 426	
2406394	2406395	1962	413	Stage I
2406394	2406395	1962	413	300H (Heads)
2406394	2406395	1963	413	300J (Heads)
2406394	2406395	1964	413	300K (Heads)
2406766	1737692	1963-71	383	
2406766	1737692	1972-74	400	
2406770	1851535	1964-65	413 426	STD
2406770	1851535	1966-69	440	Exc 69 1/2 6bbl.
2406770	1851535	1970-74	440	Exc Hi Performance
2406885	2406886	1963	426	Stage II
2406885	2406886	1964	426	Stage III
2532709	NC#	1964-65	426 Race Hemi	1.09-in. W/O Bush
2780531	NC#	1966-71	Street Hemi	1.03-in. W/ Bush
2951906	2951908	1970-74	440	All HP & HP 2
3462426	1851535	'69 1/2	440	6bbl.*
	3512061	1972	BALL STUD	Hemi

* Same as 440 exc: 100% Magnaglo inspected. The 1969 440-6 and the 1970-72 440-6 connecting rod are not the same. Hence, the 1969 440-6 does not have Six-Pack rods.

products for this engine are superior and readily available.

The stock Hemi rod has a center-to-center length of 6.86-in. and uses a 1.03-in. diameter full-floating pin. The connecting rod cap bolts are 7/16-in.

L to R:* 426 Hemi cross-bolted main bearing cap; 383-400 main bearing cap; 383-400 rear main bearing cap. *Photo courtesy* Chrysler Power *magazine.

Rod Interchangeability

Obviously, we have a great deal of interchangeability here. The problem lies not in size but in balance. If you refer to the crankshaft section you will see some cranks are internally balanced and others are externally balanced. Much of this balancing relates to the weight of the rod. Therefore, consideration must be given to matching the rods with the crank with regards to balancing.

Externally balanced crankshafts can be internally balanced with the use of Mallory metal. This is a machine shop process whereby heavy metal is added to the crankshaft counterbalances to compensate for the weight of rods and pistons.

The above discussion refers to "trick" setups for high performance track and

The dome shape of these pistons identify them as 426 Hemis. They are BRC Racing Pistons developing 12.5:1 compression.

street machines. Keep your life simple and the job inexpensive. Select a rod that matches the piston-pin-crankshaft setup in your engine.

PISTONS

All "B/RB" pistons are cast aluminum with internal steel reinforcing struts. The greatest differences between pistons are their overall dimensions, particularly bore size. Piston interchangeability, therefore, is very limited. Within the aftermarket there is such a wide variety of sizes and shapes for different types of performance—from stock to race—that it's cheaper and more cost effective to simply buy the piston you need for the performance you want.

There are three terms you'll find associated with piston description: valve relief, dome and dish. Valve reliefs are cutouts in the top of a piston to provide clearance for an open intake or exhaust valve when the piston is near TDC. A dome (pop-up) on the top of a piston increases an engine's compression ratio. A dish is a large cutout to reduce compression.

At TDC the volume between the piston and cylinder head is called the clearance volume (CV). Clearance volume includes the combustion chamber volume in the cylinder head, volume created by the head gasket spacing, the head above the block-deck surface, additional volume created by piston-to-deck clearance and by the shape of the piston top.

A dished piston adds clearance volume and a domed piston reduces it. Of these four factors, cylinder head combustion

PISTON RING SET

All ring sets have moly-coated top ring with both low (race) and medium (performance) tension oil expanders. Low tension (race) is recommended for Super Stock and all-out racing applications. Medium tension (performance) is for bracket racing use. Use only the performance rings with stock pistons. Performance = 5/64". Race = 1/16".

383

P4349853	Performance for 4.280" bore, .030" oversize
P4349854	Performance for 4.310" bore, .060" oversize
P4349831	Race for 4.280" bore, .030" oversize, use with piston
P3690828	
P4349832	Race for 4.310" bore, .060" oversize, use with piston
P3690829	

400

P4349855	Performance for 4.370" bore, .030" oversize

440

P4349856	Performance for 4.320" bore, standard size
P4349857	Performance for 4.350" bore, .030" oversize
P4349858	Performance for 4.380" bore, .060" oversize
P4349835	Race for 4.320" bore, .030" oversize
P4349836	Race for 4.350" bore, .030" oversize, use with piston P3690834
P4349837	Race for 4.380" bore, .060" oversize, use with piston P3690835
P4452994	Race for 4.4375" bore
P4529106	Race for 4.4375" bore, use in aluminum race block

Cast Aluminum Pistons for Lower Compression

P4529425	383 engines, 4.28" bore (.030" over)
P4529426	383 engines, 4.31" bore (.060" over)
P4529427	400 engines, 4.37" bore (.030" over)
P4529428	400 engines, 4.40" bore (.060" over)
P4529429	440 engines, 4.35" bore (.030" over)
P4529430	440 engines, 4.38" bore (.060" over)

Chrysler muscle cars of the '60's and '70's were designed with compression ratios in excess of 10.0:1. Today's low octane, unleaded fuels will not perform satisfactorily in these high-compression engines. You can lower the compression of your engine to 9.0:1 with the use of the above cast aluminum pistons. Or, you can increase performance in engines that already have a lower compression ratio. Either way, you must use the performance (5/64") ring set.

chamber volume is usually the biggest contributor to clearance volume.

MATCHING CRANK, RODS & PISTONS

Keeping the above in mind, let's look at some engine design basics. Consider different crankshaft, connecting rod and piston combinations. As stated earlier, some of these parts are interchangeable. But with any interchange, the manufacturer's specifications for the engine must be adhered to religiously. Otherwise, some expensive damage will result.

For each engine there is a specific *crankshaft throw, connecting rod center-to-center length, piston compression height* and *deck clearance.* If any of these dimensions are changed, compression will be lowered or raised and engine performance will be affected—possibly to the extent of the pistons hitting the cylinder heads. Let's see how this works.

With a given crankshaft, if you install a longer rod, the piston will operate higher in its bore. Conversely, a shorter rod will move the piston lower in its bore.

Another term identified with compression is piston compression height. This is the distance from the piston-pin-bore center to the flat surface around the top of the piston.

When compression height is increased, the distance between piston and pin and the piston top is increased. Therefore, the top of the piston operates higher in its bore. Reducing compression height lowers where the piston operates.

Our final consideration is deck clearance, that is, the distance from the top of the piston at TDC and the block's deck surface. These specifications must be kept in mind when considering a swap.

Never arbitrarily change crankshaft throw, connecting-rod length, piston compression height or deck clearance. Without careful consideration of the

"B" ENGINE CYLINDER HEAD CASTING NUMBERS

Year	Engine	Casting Number
1960-62	361/383/413	2206324
1962	413 Max Wedge	2402286
1964-67	361-383	2406516
1964	426 Max Wedge	2406518
1963	361/383/413	2463200
1963	426 Max Wedge	2463209
1967	440 High Perf.	2780915
1968-70	383-440	2843906
1971-72	383/400/440	3462346
1973	400-440	3462346
1973	400-440 Motor Home	3751213
1974	400-440	3769902
1975	400-440	3769975
1976-78	400-440	4006452
Stage IV		3614476
Stage V		4286526V

above measurements you could ruin an engine.

Remember our previous discussion of balance in the rod section. As a general rule, any interchange of lower-end parts should be balanced professionally at your local machine shop.

CYLINDER HEADS

All "B" and "RB" heads are interchangeable. The only problem arises in trying to use 413-426 Max Wedge heads on a "B" (383-400) because there is no intake manifold available for this combination.

The history of the "B" head goes back to 1958. Up to 1961, the "B" engine used a head with 1.95-in. intake valves and 1.60-in. exhaust valves. The heads (up to 1963) can be identified by their aluminum rocker-shaft attaching brackets and valve covers retained with only four bolts.

In 1962, the intake valves were increased to 2.08-in. on the intake but the smaller 1.60-in. exhaust valves were retained. With the introduction of the now-famous 300J head, the exhaust valves were increased in size to 1.74-in. and the 6-bolt valve cover became the standard.

The "modern" "B" engine head that was produced from 1964 to 1966 retained the 2.08-in. intake valve but reduced the exhaust valve back to 1.60-in. This head was quite similar to the 1962 design, but continued the 6-bolt valve cover.

With the introduction of the Max Wedge series in 1962 (413) and the 1963-64 426, a much larger exhaust valve (1.88-in.) was incorporated. Additionally, these heads had very large ports and no heat cross-over passage. Many of the Max Wedge heads now incorporated adjustable rocker arms, where previously, only non-adjustable rockers were used.

In 1967, the 440 High Performance head was introduced, returning to the 1.74-in. exhaust valve. This head superseded the old 300J. In 1968, a larger combustion chambered version of the 440 High Performance head was brought to market and was manufactured until 1971.

In 1972, to meet new emission standards, a flatter intake port was introduced on all "B" series engines. This can be identified by casting number 3462345. In mid-1973, this head was further refined by adding extra cooling passages, making it a bit more durable. These heads remained the same through 1978, with only a few changes such as the addition of hardened valve seats to overcome the effects of unleaded gas.

Today, an active aftermarket exists in

These are the top and bottom views of the 426 Hemi head. You can easily see the hemispherical shape of the combustion chamber.

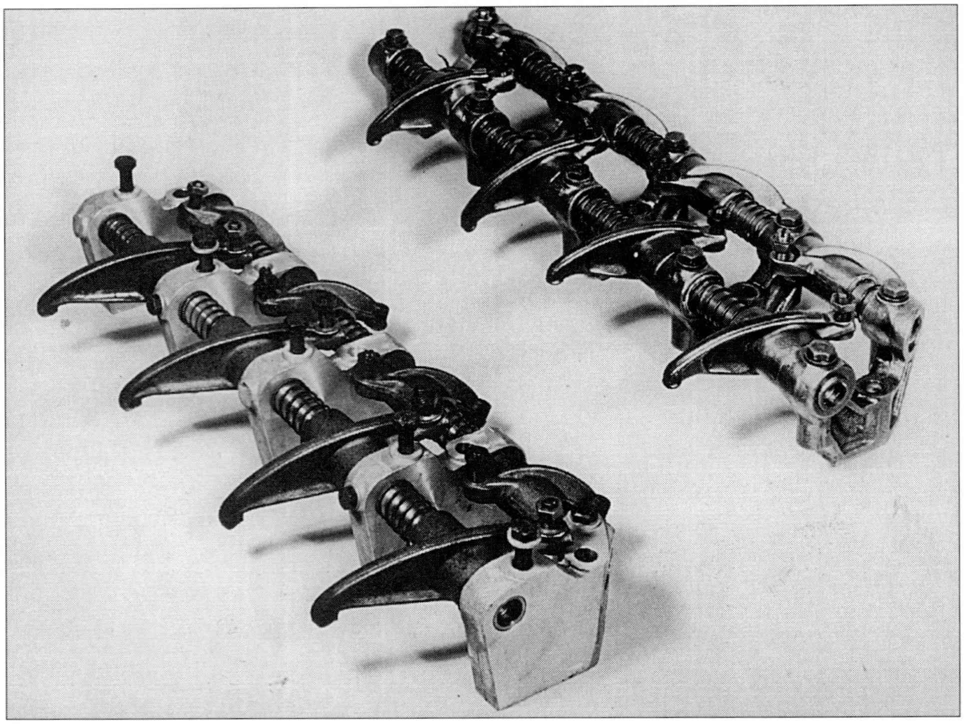

At right is the rocker arm assembly of a stock Hemi. At left is a high performance aftermarket unit.

highly refined heads designated Stage IV, V or VI. These are high performance heads with revised port shape, improved combustion chambers and oversized valves. Consider these only if you're into high performance or racing.

Page 57 has a complete list of casting numbers to help you identify your head or the one you may be looking for.

426 Hemispherical Chambered Heads

In 1964, Chrysler resurrected the defunct 1951-58 Hemi in two racing versions: one for oval track racing and one for drag racing. Neither of these engines was suitable for street use. In fact, Chrysler cautioned users of the drag-tuned engine against running it wide open for more than 15 seconds. The 1966, 426 cid was simply a de-tuned version of the '64 racing engine.

This detuning was achieved, in part, by reducing the compression ratio of 11:1 and 12.5:1 on the racing engine to a more "street friendly" 10.25:1. By today's standards and low octane gasoline, a 10:1 compression ratio is hardly street friendly. But in 1966 times were different.

The original Hemi heads (1964) were cast iron with an almost hemispherical-shaped combustion chamber. This separates it from its counterpart 426 wedge-shaped combustion chambered head. The 1965 Super Stock heads were aluminum. The 1966 through 1971 race Hemis were cast iron. All of these heads are interchangeable and the same intake manifold will fit both cast-iron and aluminum heads.

The intake valves are 2.25-in. and the exhaust are 1.94-in. The spark plug hole is located as close to top center of the chamber as could be made. The most notable exterior difference between the wedge and Hemi are the dual, adjustable rocker arms.

As with the wedge heads, there are a number of aftermarket Hemi heads available. Most of these are aluminum with increased breathing capabilities, bigger valves and some with dual spark plugs. Stage V Engineering of Walnut, California, makes a full-on aluminum Hemi head that can be bolted (after slight modification) to any 383 through 440 wedge!

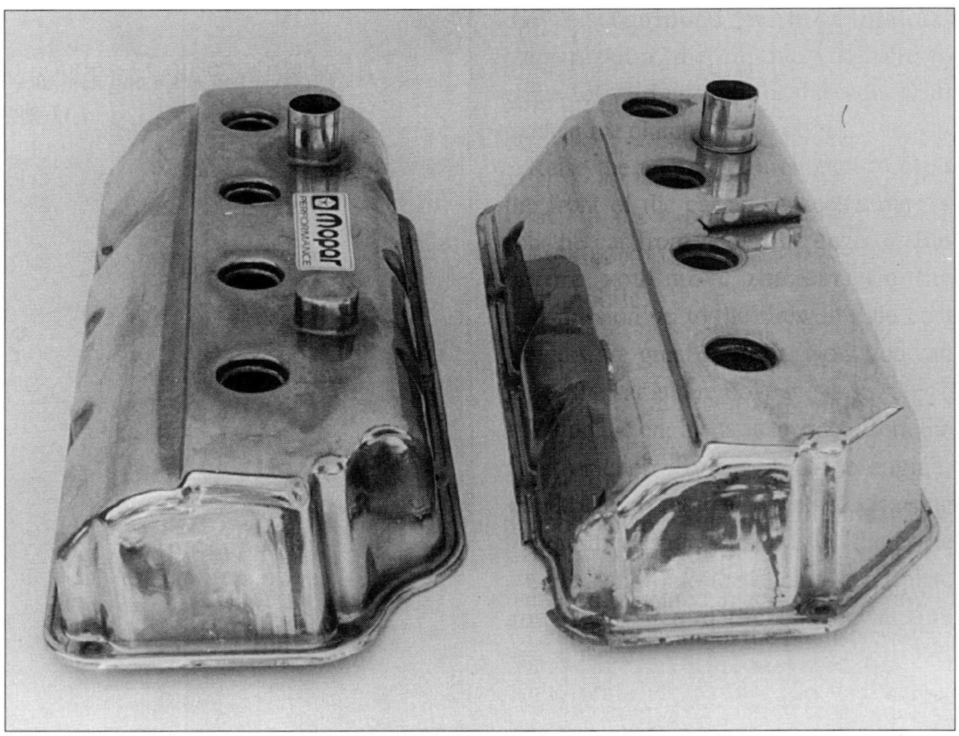

There are quite a number of street and racing Hemi heads around—many with different configurations. The valve cover on the right is a stock unit for 1966 and later. The cover on the left is for a K-head. Compare the curved cutout of the performance unit compared to the flat bevel of the stock cover.

INTAKE MANIFOLD CASTING NUMBERS

Part #	Casting#	Year	Engine	Remarks
	1851898	1959-60	383 1-4bbl	
	1851899	1959-60	383 1-2bbl	
1854817	2264877	1959-60	413 2-4bbl	300E-F-G
1854852	1827899	1959-61	383 2-4	343hp
2240394	1947162	1961	383 & 413	2-4bbl Left
2240395	1947163	1961	383 & 413	2-4bbl Right
1859300	2205968	1962	383 1-4	
1859852	1827899	1962	383 2-4	343hp
2402332	2206000	1962	413 1-4	
1858623	2206000	1962	413 1-4	
1854817	2264877	1962	413 2-4	385hp
1854817	2264877	1962	413 2-4	300H
	2402726	1962	413 2-4	Max Perf I
1859300	2205968	1963	383 1-4	
1858623	2206000	1963	413 1-4	Exc 300J
2421368	2129985	1963	413 2-4	300J Right
2421369	2129987	1963	413 2-4	300J Left
	2402726	1963	426 2-4	Max Perf I
2465214	2402728	1963	426 2-4	Max Perf II
1859300	2205968	1964	383 1-4	
1858623	2206000	1964	413 1-4	
2421368	2129985	1964	413 2-4	300K Right
2421369	2129987	1964	413 2-4	300K Left
2465214	2402720	1964	426 2-4	Max Perf III
	2465165	1964	426 1-4	Max 2-PL Short
	2465185	1964	426 1-4	Max 1-PL Short
2468418	2468045	1964	426 1-4	S/S Hemi Strip
1858623	2206000	1964	426 1-4	Street Wedge
	2468043	1964	426 1-4	Hemi NAS-Track
1859300	2205968	1965	383 1-4	
1859300	2205968	1965	383 1-4	
1858623	2206000	1965	413 1-4	
2468418	2536900	1965	426 2-4	S/S Hemi Strip
1858623	2206000	1965	426 1-4	Street Wedge
	2531631	1965	426 1-4	Hemi/Holley
	2946278	1965	426 2-4	Hemi/Holley
1859300	2205968	1966	383 1-4	
	2531921	1966	426 2-4	Hemi/Holley
2780542	2780543			
and	2780544	1966	426 2-4	Hemi/Carter
1858623	2206000	1966	440 1-4	
1859300	2205968	1967	383 1-4	
2780542	2780543			
and	2780544	1967	426 2-4	Early Hemi/Carter

INTAKE MANIFOLD CASTING NUMBERS (continued)

Part #	Casting#	Year	Engine	Remarks
2531943	2780543			
and	2780544	1967	426 2-4	Hemi S/S
2780542	2780543			
and	2780544	1967	426 2-4	Hemi/Carter
2468418	2536900	1967	426 2-4	Hemi NASCAR
2843031	2806178	1967	440 1-4	HP
2843684	2806301	1968	383 1-4	
2780542	2780543			
and	2780544	1968	426 2-4	Hemi/Carter
248418	2536900	1968	426 2-4	Hemi/Holley
2843031	2806178	1968	440 1-4	All
2843684	2806301	1969	383 1-4	
2780542	2780543			
and	2780544	1969	426 2-4	Hemi
2843031	2806178	1969	440 1-4	
3412046	3412046	1969	440 3-2	Edelbrock
2951665	2951666	1970	383 1-4	
2780542	2780543			
and	2780544	1970	426 2-4	Hemi
2951737	2951736	1970	440 1-4	
2946275	2946275	1970	440 3-2	Early
2946275	2946276	1970	440 3-2	
3412046	3412046	1970	440 3-2	Edelbrock
2951665	2951666	1971	383 1-4	
2780542	2780543			
and	2780544	1971	426 2-4	Hemi
3512502	3512501	1971	440 1-4	
3577164		1971	440 1-4	Imperial
2946275	2946276	1971	440 3-2	
3614047	3614046	1972	400 1-4	W/O Calif Emmission
3614048		1972	400 1-4	With Calif Emission
3614015	3614014	1972	440 1-4	W/O Calif Emission
3614016		1972	440 1-4	With Calif Emission
3614017		1972	440 1-4	W/O Calif Emission
3614018		1972	440 1-4	With Calif Emission
2946275	2946276	1972	440 3-2	Early
3671880	3671879	1973	400 104	
3698585	3698444	1973	440 1-4	
3698586		1973	440 1-4	Imperial
3698584	3698442	1974	400 1-4	
3751729	3751728	1974	440 1-4	

Two views of an inline, dual 4-barrel Hemi intake manifold for street use. Note the casting numbers on the end view.

Design Fundamentals

To further understand the effects of head swapping, let's look at some design fundamentals. Three terms should be defined. These are top dead center (TDC), bottom dead center (BDC) and swept volume (SV).

TDC is the highest position the piston reaches in its upward travel. BDC is the opposite, or lowest position reached.

As a piston travels from BDC to TDC, it sweeps through, or displaces, a volume called *swept volume*. Swept volume is a cylinder's displacement.

A cylinder's compression ratio is directly proportional to its swept volume or displacement, and is inversely proportional to its clearance volume. Therefore, we can make a formula that states:

$$\text{Compression Ratio} = \frac{\text{Swept Volume}}{\text{Cylinder Volume}} + 1$$

This compression ratio formula says that as an engine's displacement is increased, clearance volume must also be increased to maintain the same compression ratio. That's why oversized pistons are manufactured to increase clearance volume by a smaller-than-original dome or larger-than-original dish. Otherwise compression ratio would be increased.

This is an important consideration when interchanging heads. With today's fuel, compression ratios over 8:1 can cause detonation. If detonation is severe, engine damage will occur. With stock high-compression engines (10.25:1), this is already a problem. At the same time, changing to a head with a combustion chamber volume that is too large will result in a low-compression, poor-performing engine.

Valve & Port Size—Other head design factors affecting performance are valve and port sizes. If these are too small, the engine won't breathe properly and will lose power. In other words, the engine cannot draw an adequate air/fuel mixture through the smaller valve and

Contrast the two views of the street intake above with the two views of this single 4-barrel NASCAR intake.

port openings. On the exhaust stroke it will be unable to expel the exhaust efficiently.

If the valves and ports are too large—particularly on the intake side—the air/fuel mixture will not have enough velocity to keep the fuel mixed with air at low rpm. This creates poor driveability and low-speed sluggishness.

There is a good rule-of-thumb: with an otherwise-stock engine, keep the valve sizes within +/- 10% of engine specifications.

IGNITION

All of the "B" engine distributors are interchangeable with one another as are the "RBs." Because of the raised block (RB), the distributor shaft for that block is longer than its "B" counterpart. If you wish to use an "RB" distributor in a "B" block, Chrysler has accommodated you by providing an aluminum spacer (PN P3690275). This also allows the use of a 426 Hemi distributor with the "B" series.

Within their classifications, and based on block series, single-point, dual-point and electronic distributors were used with the 383 through 440 engines. Single-point and electronic distributor housings are cast aluminum while the dual-point is cast iron.

Be aware that distributors are calibrated differently for specific engines,

WATER PUMP HOUSING

Part #	Casting#	Year	Engine	Remarks
1944916		1962	383, 413	W/O A/C
2240318		1962	383, 413	With A/C
2240318		1963	383, 413	With A/C
2240318		1963	413	300J Head
2448758	2205862	1964	413	300K Head
2448758	2205862	1964	383, 413, 426	With A/C
2448759		1963-64	383, 413, 426	W/O A/C Exc
2525498		1965	383, 413, 426	W/O A/C
2525498		1966	383, 413, 426 440	W/O A/C
2536086	2263317	Parts	383, 413, 426, 440	Aluminum
2536725	2205862			
	2536767	1965	426	Alum S/S Hemi
2585314	2468007	1965-66	383, 413, 426, 440	With A/C
2585314	2468007	1965	426	Hemi/Steel
2585314	2468007	1966	426	Hemi
2780988	2780987	1967	383	
2780988	2780987	1967	426	Hemi
2780988	2780987	1967-68	383, 440	With A/C&HP
2780988	2780987	1967-68	383, 440	W/O A/C
2780988	2780987	1968	383	
2780988	2780987	1968-69	426	Hemi
2780988	2780987	1969	383, 440	
2780988	2780987	1971	383, 426, 440	
2951806		1970	383, 426, 440	
3671315		1972	400, 440	
3751186		1973	400, 440	
3769192	3698468	1974	400, 440	

Looking for that ultimate edge? This Hemi intake has been custom-fabricated. Note the weld lines.

Side views of the late high-volume pump and the non-air conditioning or low-volume pump.

even though they are mechanically interchangeable and appear the same. Consequently, if you arbitrarily change distributors or distributor components, your engine's performance may be altered. Power will be reduced if total spark advance is not sufficient or timed right, or you may end up with a detonating engine if there's too much advance. I suggest consulting a good ignition specialist to help you select or modify a distributor other than stock.

Consider the differences when swapping distributors. Remember, breaker-point and electronic distributors are interchangeable only if the complete ignition system is changed.

WATER PUMPS

All "B" engine and Hemi water pumps are interchangeable. An aluminum water pump housing is available which will fit all "B" and "RB" engines. Problems arise, however, when you begin to deal with air conditioning.

All pre-1972 engines ("B" and "RB") used a different water pump on engines with air conditioning. In 1973 this was changed to a standardized A/C water pump. For those interested in a bit of horsepower gain, the air conditioning water pumps require less horsepower to drive than do the non-A/C water pumps.

Mopar Performance notes that the alternator bracket attaching hole is not in the same place on the aluminum water pump housing as it is on the 1968-75 housings. To correct this, a pre-'68 alternator bracket should be used with the aluminum housing. ■

***L to R:* Early model high-volume water pump impellers; late-model high-volume water pump; non-air conditioning water pump impellers.**

RECONDITIONING THE SHORT BLOCK

5

Sure, it looks scary! You've got parts everywhere (but of course you've marked them so they'll return to the correct place), the bare block is sitting there staring at you and you're sure I've led you down the Path of Fools. Well, fear not; neither of us are fools and we will indeed get this engine back together. In fact, if you're patient, use lots of common sense and the information I've provided, the engine will go back together better than new.

In this chapter I'll show you how to inspect, clean and recondition the parts. Additionally, we'll visit a machine shop to see how all your parts will be machined to make them even better than they were when they came from the factory. Pay attention and read the charts and graphs carefully. Use the information and your best judgment and you'll soon be assembling your new short block.

CLEANING & INSPECTION

In every chapter I stress cleanliness. I must stress it here and I'll emphasize it again in the following chapters. You just can't get the parts too clean. One metal filing between a bearing and a rotating surface and both parts will be destroyed. This could be your crankshaft or your cam—neither of which you will want to replace because you were lazy or careless when it came to cleaning. Once the part is clean, keep it clean. Heavy-duty trash bags are excellent engine covers and

No, this isn't how the block looked after it came out of the car, but I thought I'd give you a glimpse of where we're headed. We've much to do before getting to this point.

Make your life easier by getting a can of Gasket Remover, then saturate the old gasket and wait about 10 minutes for it to penetrate. Now is a good time to get another cup of coffee.

containers for clean parts.

Before you jump in and start cleaning, however, check the head gaskets and deck surfaces for signs of leaks.

Head Gaskets & Block Deck Surfaces

Take a close look at the head gaskets, particularly if the engine had compression or cooling system problems. Coolant leaks show up as rust streaks on the block deck or cylinder head gasket surfaces.

Compression leaks appear as gray streaks radiating from a combustion chamber or cylinder. A compression leak between a combustion chamber and water passage will over-pressurize the cooling system. This causes the engine to lose coolant.

A compression leak to the outside or between cylinders would only have been found with a compression or leak-down test. If you discover signs of a leak, check the block and heads for warp or other imperfections. Checking the block's deck surfaces begins on page 68.

A serious deck-surface problem to check for is *notching*. Notching occurs when hot exhaust gases remove metal, much in the same manner as an acetylene torch. Notching is more prevalent with racing engines. It also occurs when any engine is driven a long time with a blown head gasket. Check the severity of any notches after cleaning the block's deck surface.

Gasket Removal

Although hot tanking will clean the gasket surfaces, it's best to do this yourself. This gives you a chance to check the surfaces before the block is carted off to the machine shop.

A gasket scraper will save a lot of time, skinned knuckles and difficulty, and is better than a screwdriver or putty knife.

The easiest way to remove cooked-on gaskets is with the aid of an aerosol gasket remover. This can be purchased at any auto parts store. Spray the remover onto the gasket surface, fully saturating it. Allow it to work its way down through the gasket material and break the bond between gasket and metal. Now your scraping job will be much easier.

Scrape all the gasket surfaces—both cylinder deck, front-cover, oil pan and intake manifold. It's best to clean all the gasket surfaces at one time, so scrape the heads also.

Getting It Clean

If the block is going off to the machine shop to have extensive work done, as a matter of course the machine shop will hot tank the block. This will be either the chemical operation or the thermal type. A

After about 10 minutes, the gasket should come off fairly easy with a scraper.

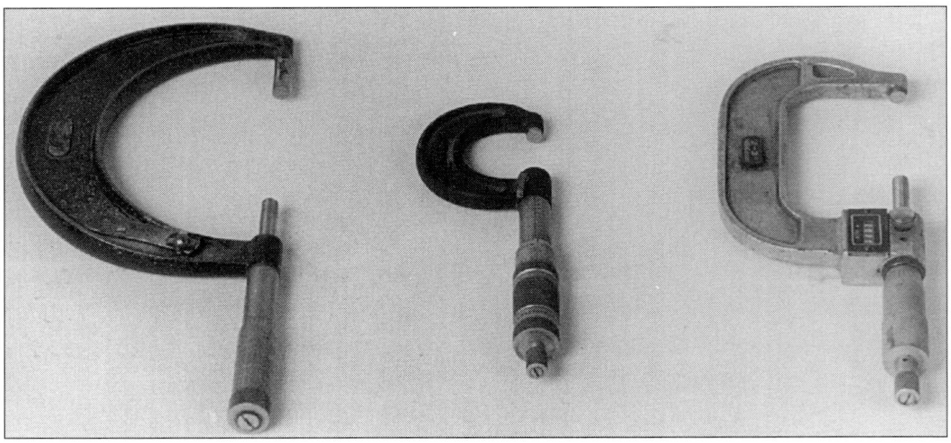

Throughout this and the next chapter you're going to need a number of micrometers. Borrow them if this is the only engine work you'll be doing. If you plan to do lots of engine work, buy a full set.

As before, sand all the gasket surfaces with 80-grit sandpaper.

I'm trying to open the coolant passages with my rat-tail file. Do you remember the photo of the corroded closed thermostat in Chapter 2? This whole block is clogged with rust. I'm sure it never saw a dime's worth of antifreeze!

good machine shop will be as meticulous about cleanliness as you are. If you want to save your cam bearings, as most people do, then the machine shop will want to wash the block in their cold tank. Sometimes, however, people just like to do it all themselves.

If this is the case, or if you don't plan to have any machine work done, block cleaning can be done at home. So, if this is what you decide to do, here's how to do it.

Buy a gallon or two of water-soluble degreaser. Take the degreaser, some stiff brushes, your oil-gallery (or rifle cleaning) brushes, some pipe cleaners, mechanic's wire and a few rags along with the block to the local coin-operated car wash. Be sure they permit washing greasy things like engine blocks—most do. I suggest the car wash rather than your backyard because of environmental concerns regarding the disposal of the oil, grease and such. Many communities have laws against this, and you could very well be fined heavily if spotted cleaning your block in the driveway. Just be careful and check with your local environmental protection people. Now, let's get back to work.

Slosh on the degreaser. Coat all the surfaces thoroughly and pour some into the oil galleries. Let it work for a while. If it begins to dry, apply more. You may even want to scrape a little here and there and then apply more degreaser. For really tough places try a liberal application of carburetor cleaner or Barryman's B-12.

With a very stiff brush, work the degreaser in. Get down into all the holes, crevices, galleries, openings, etc. Now, use the high pressure rinse cycle of the wash hose to blast off loosened grease and grime.

You'll find areas that need to be redone, so go over them again. When the exterior is as clean as you can get it—all the crevices, nooks and crannies—go after the interior, particularly the oil galleries and dozens of holes. Use any or all of the brushes you brought along: bottle brushes, pipe cleaners, gun brushes, acid swabs or anything to clean the holes and galleries. Use the mechanic's wire to pull rags through them. This should give you a really clean block, ensuring a professional result.

Thread Chasing

Although the block should now be as clean as a whistle, there are still a few chores left. Chase the threads in all the bolt holes. This is done with a bottom tap—not a taper tap.

The bottom tap, as its name implies, cuts or cleans the threads all the way to the bottom of the hole. A taper tap, used to start threads, doesn't have full cutting threads at its end and will not do a complete job.

Thread chasing serves three purposes: it makes it much easier to install bolts, it limits the possibility of cross-threading and it ensures accurate torquing.

Pay particular attention to the head bolt and main bearing cap bolt holes. For this job use a 1/2-13 tap—1/2-in. diameter with 13 threads per inch. This is also designated 1/2 NC. NC stands for National Coarse. (This distinguishes between coarse and fine threads with NF, the designation for National Fine.) Another word of caution: the tap should start fairly easily. If it binds, remove and check it. You don't want to cut new threads, just clean the existing ones. Turn the block over so the debris can fall from the threaded holes you just cleaned.

The final cleaning detail is to ensure the water passages are free of loose rust, deposits and core sand. Scrape out the passages and force water through them. Use your bottle brush or anything else to

Chase the threads for the head bolts no matter how lazy you get with the rest of them. This will help insure accurate torque adjustments preventing future blown head gaskets.

get them clean. Pay close attention to the passages connecting the cylinder heads to the block to ensure a good flow of water between block and heads. Use a rat-tail file to clean these. Be careful, however, not to scratch the deck surfaces. This could cause a head gasket leak.

If you have an air compressor, use an air gun to blow out hard-to-wash areas. This will help dislodge stubborn dirt and grease and will dry remaining moisture.

Moisture is one of the big bugaboos that leads to problems. Even humidity will cause rust, so coat all bare metal surfaces with oil. I like the spray-on, water dispersing types such as WD-40. This is a very light oil, easy to work with and the spray gets into areas otherwise difficult to reach. Whatever type of oil you use—spray, oil-can or wipe-on—be sure that all machined surfaces are well oiled.

FINAL BLOCK INSPECTION & RECONDITIONING

The next step in reconditioning the block is to determine just how extensive the project will be. The following processes will determine if the cylinders require boring and, if so, how much. If the head gaskets indicated problems, you will be able to determine whether or not the deck needs resurfacing. Finally, you'll check all of the machined surfaces to determine if there is excessive wear.

To perform these checks you'll need the following: 3-4 in. inside and outside micrometers, telescoping gauges, a machinist's straight-edge and a set of feeler gauges. You'll also need a 2-3-in. outside mike to check the crankshaft main and rod-bearing journals.

Bore Wear

Bore wear dictates whether you can simply hone the cylinders or if you must bore them and fit new pistons.

Cylinder bores don't wear the same from top to bottom. The majority of bore wear is confined to the top 1-in. of compression-ring travel.

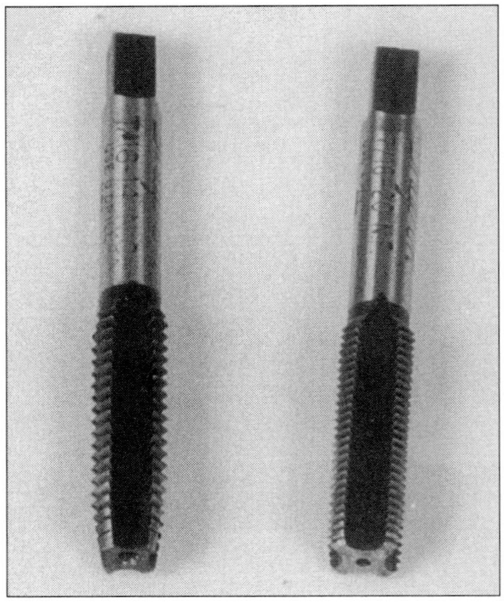

The tap on the left is a starting or taper tap. The tap on the right is the tap of choice for the job—a bottom tap. Notice that it has cutting threads all the way to the bottom.

Combustion and compression pressures are highest at the top of the stroke and decrease as the piston travels toward BDC (Bottom Dead Center). This pressure forces the compression rings against the cylinder walls, causing more wear at the top of the stroke. Toward the bottom of the bore the rings are relatively unloaded and better lubricated, causing little wear. The result is an actual taper, or cone shape, within the cylinder.

Look carefully at the lower part of the bores. You should still be able to see the cross-hatching or honing marks. The top of the cylinder will be shiny or glazed.

You'll also discover a bore doesn't wear evenly around its circumference. One bore will wear differently from another. This is due to cylinder-wall temperature differences. See the graph on this page.

Measuring Bore Wear—Because a bore wears little at the bottom, the difference between wear at the top, just below the ridge, and the bottom of the bore is the amount of taper. Measure the bore parallel to the centerline of the engine and then 90° to the centerline. Make the measurement in two or three

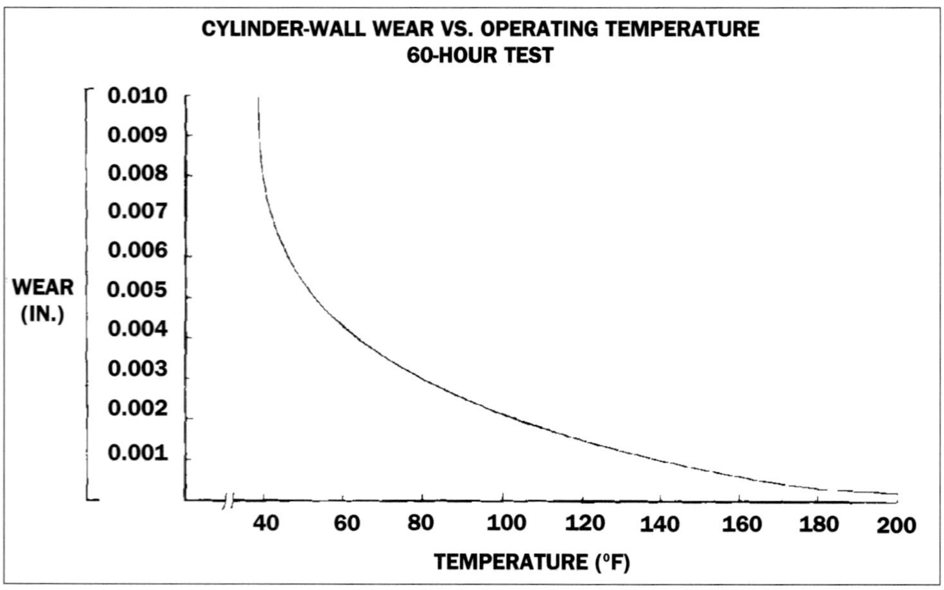

Bore wear increases as operating temperature goes down. Never run an engine without the thermostat in order to lower operating temperatures. Solve temperature problems and then use a thermostat of the correct operating temperature range.

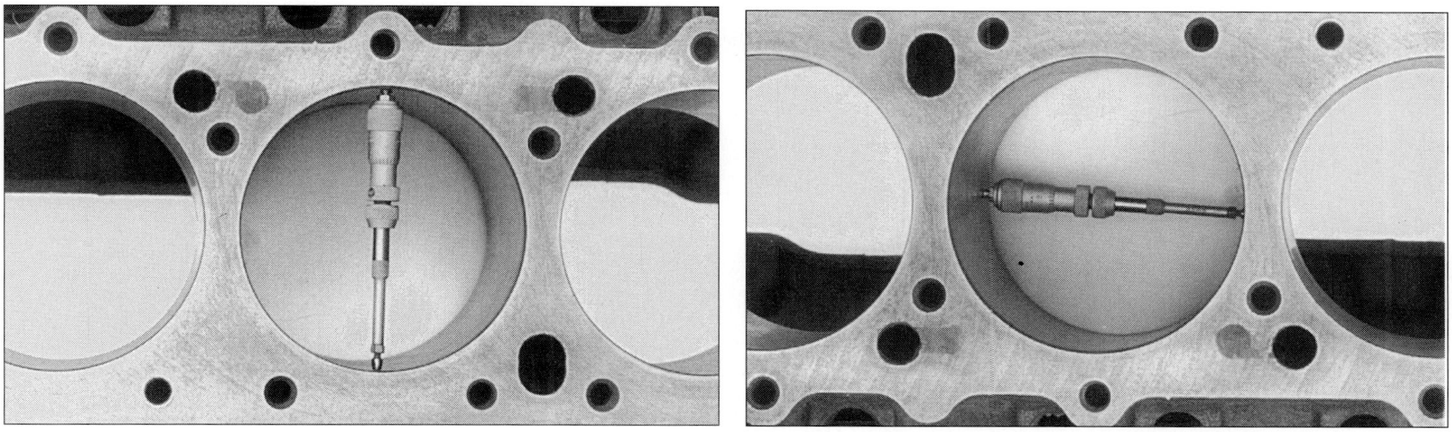

By using an inside micrometer you can easily determine bore taper. Take the first measurement just below the ridge. This is the area of maximum wear. Take the second reading at the bottom of the bore where there is little or no wear. The difference is the amount of taper. The same job can be done with a telescoping gauge and outside micrometer.

other areas of the bore and use the highest figure. This will determine what and how much will have to be done to that cylinder to restore it.

Sleeving—Each cylinder will have to be measured, because the cylinder with the greatest wear is your guideline to determine the extent to which all of the cylinders will have to be reworked. There's one exception to this. If one cylinder is cracked or worn beyond salvaging and the other seven are okay, then the offending cylinder can be *sleeved.* If more than one cylinder needs sleeving, it might be more economical to junk the block and buy another. Check prices first.

Although it is a common repair, I feel sleeving a cylinder is a pretty "iffy" situation. I have seen more than one sleeve "pop out," forcing the owner to start all over from scratch. Usually, the machine shop will only guarantee the sleeve for 90 days and often, only if they do all the assembly work on the engine. Usually, a junked block is just as cheap and will assure you many thousands of worry, and trouble-free miles. On the other hand, sleeving may be your only option if you're restoring a rare Mopar with a 426 Hemi, and need to keep all numbers "matched."

As you measure the bore taper, you may find the two front cylinders and two rear have greater wear than the center four. This would be due to the phenomenon of greater wear at lower temperatures.

The graph on page 68 illustrates the rapid increase in wear as the operating temperature drops below 180°F. Because of the wear pattern in the end cylinders, wear measured parallel to the crankshaft should exceed that measured 90° to the crankshaft.

Even though your decision to bore or hone will be based on the cylinder with the most wear, this won't provide a final figure as to how much the cylinder will have to be bored.

Uneven wear shifts a bore's centerline in the direction of maximum wear, giving the bore an irregular shape. To get the bore back to its original centerline usually requires removing more metal than indicated by the taper measurement.

The machinist will decide how much metal must be removed to clean up the bore. Starting with a 0.010-in. oversize bore, he or she will move up in 0.010-in. increments until the bore cleans up the entire length. They will start with the bore with the worst wear/taper, because this will establish the minimum size for all the bores.

Many quality machine shops with modern equipment will eliminate this step-by-step increase, measuring the bore with the greatest amount of wear then machining all the bores to the nearest 0.010-in.—0.010, .020, .030 or .040 oversize. A word of caution follows, however. The casting on 1976, '77 and '78 blocks for the 400 and 440 are a thin-wall design. This thin-wall structure

CYLINDER BLOCK SPECIFICATIONS

	Main Bearing Bore	Cylinder Bore
318	2.625-in.	4.250-in.
400	2.625-in.	4.340-in.
413	2.750-in.	4.180-in.
426	2.750-in.	4.250-in.
426H	2.750-in.	4.250-in.
440	2.750-in.	4.320-in.

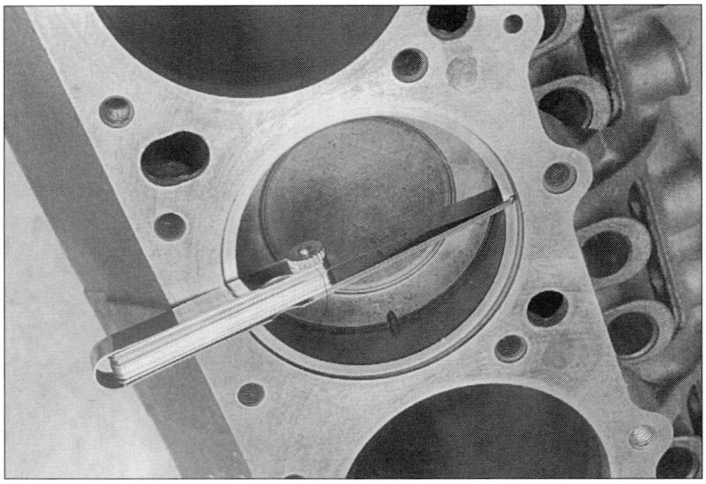

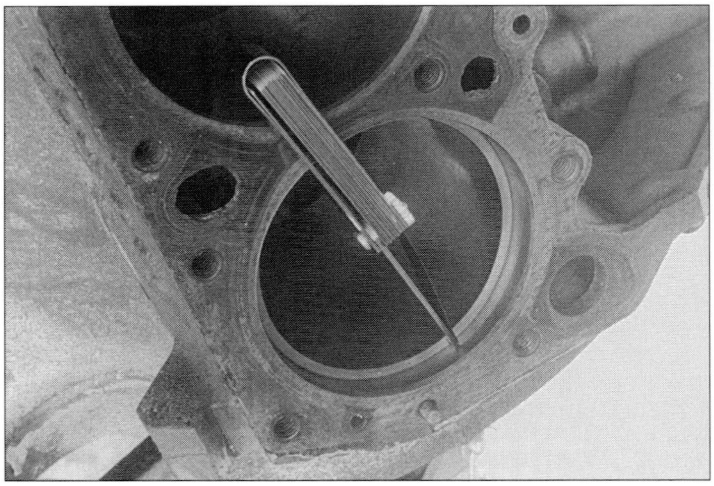

To more accurately photograph measuring ring gap I have selected the very top of the piston, rather than the bottom as described below. The second measurement is taken below the ridge. Subtracting minimum wear (top or bottom) from maximum wear (below the ridge) will give you approximate bore wear.

prohibits overboring more than 0.020-in. Refer to page 50 to determine what year your casting may be.

Three common tools are used to determine bore wear. If you have access to a dial-bore gauge, this is by far the most accurate. Short of this a 3-4-in. inside micrometer or telescoping gauge and an outside micrometer will give accurate results. If neither of these is available, use the old standby piston-ring-and-feeler-gauge method. With these two you can compare the ring end-gap at different positions in the bore. These will correspond to bore taper.

Dial-Bore Gauge or Micrometer Method—To use the dial-bore gauge or inside micrometer, measure the cylinder diameter at the bottom of the bore. Then, measure the bore just below the ridge. Take three or four readings at different points around the circumference at this point.

If you're using an inside mike, subtract the bottom figure from the largest top figure to determine the maximum amount of taper for that cylinder. With a dial-bore gauge, zero the gauge at the bottom of the bore and read the amount of wear directly as you move the gauge around the bore at the top.

Ring & Feeler Gauge Method—The ring and feeler gauge method will indicate minimum wear between the top and bottom of the cylinder. Because a ring can't conform perfectly to a worn bore, you will not be able to determine maximum wear. The more irregular the cylinder wear, the less accurate the reading will be. However, this method is sufficient to determine whether you will have to bore and fit oversize pistons. Or, it may indicate that you can get by with honing and using moly rings.

The method is quite simple. Place a compression ring in the bore just below the ridge. Push it in with a piston without rings to square it in the bore. With your feeler gauge, measure ring end gap and note it. Again, using a piston without rings, push the ring down to the bottom of the bore. Now, measure the end gap again and subtract this measurement from the top one. To determine the amount of taper, use the chart on p.66. Note that taper is approximately 0.3 times end-gap difference.

G_2-G_1 $\triangle$G	TAPER
0.000	0.0000
0.001	0.0003
0.005	0.0016
0.010	0.0032
0.015	0.0048
0.020	0.0064
0.025	0.0080
0.030	0.0095
0.035	0.0111
0.040	0.0127
0.045	0.0143
0.050	0.0159

Approximate Taper = 0.30 x $\triangle$

Using ring end gap to determine bore taper. Maximum end gap (G2) is found with ring placed immediately below ridge. Minimum gap (G1) is measured with ring in unworn section of the bore. Use the method described in the illustrations to measure end gap.

How Much Taper Is Okay?—You're the engine builder so you get to make the decision. What do you want from your rebuild? If you only need 10,000 to 20,000 miles before oil consumption starts again, then you can "get by" with as much as 0.010-in. taper using the micrometer method or 0.008-in. using the ring and feeler gauge method. If, however, you want a truly fine job that will give you 100,000 or more miles, a 0.006-in. taper with the mike and 0.004-in. with the ring and feeler gauge method should be your maximum figure. Anything over this should call for boring and new pistons.

Let's look at the reason for these figures. To seal between the piston and cylinder wall, the rings must expand and contract as the piston travels up and

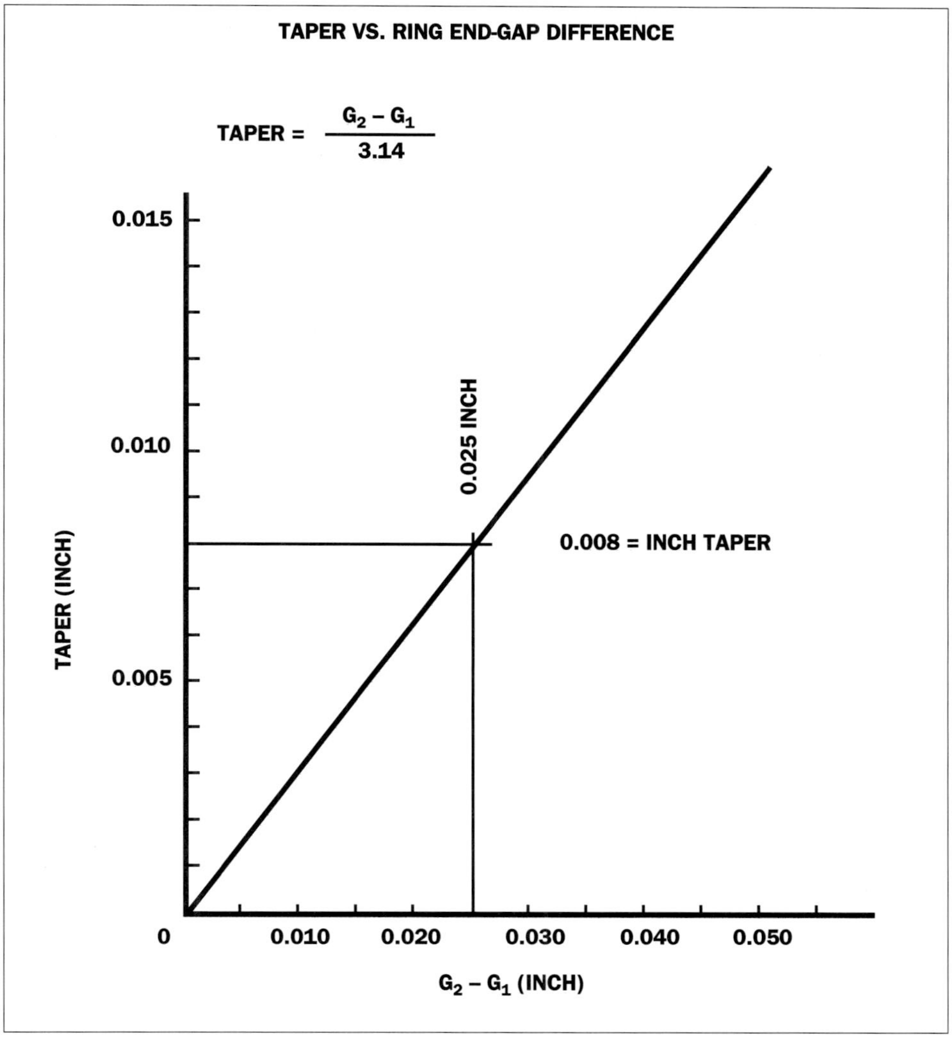

After finding the maximum difference between two end-gap readings, use this curve or chart to approximate bore wear.

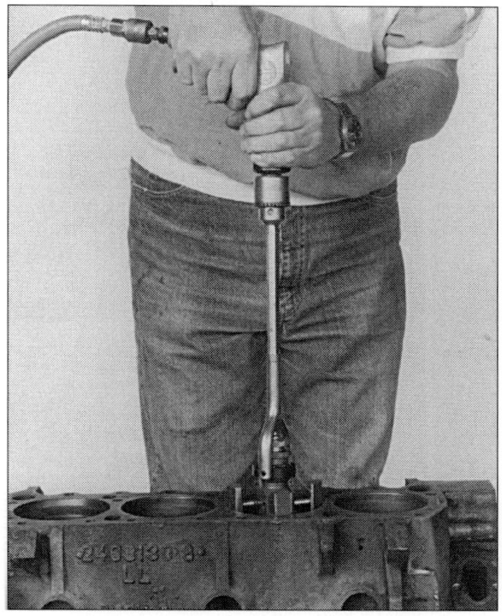

No, I'm not hand-boring the cylinders—I'm honing them with this really nice, fully adjustable honing tool.

Here is your selection of honing tools. From left to right you can see they increase in value and therefore, performance. All of them, however, will get the job done.

down. The greater the taper, the more the rings must expand and contract. This produces metal fatigue in the rings. The ring expansion and contraction wears the piston's ring lands. Soon the rings are unable to seal.

If taper is removed by boring, there is no expansion and contraction problem for the rings. This reduces ring fatigue. They continue to seal until the bore wears and new taper develops. By the time excess ring expansion and contraction begins the engine has gone another 100,000 miles.

So, if you only need to squeeze out another 10,000 miles you can get by with 0.010-in. taper. If you're in love with the car or are restoring it back to factory specifications, have the block bored if taper exceeds 0.006-in.

Piston-To-Bore Clearance

The discussion on bore taper is to answer one question: can you reuse the old pistons? If you can you'll save a ton of money (as much as $400). If the pistons are worn or damaged beyond use, taper becomes a moot point. Assuming there are no cracks or broken skirts, measure the piston-to-bore clearance. Later in this chapter we cover piston-related problems you'll want to check before making the final decision on whether to reuse the old pistons.

Be sure the block and pistons are at the same temperature—ambient temperature is fine. Aluminum and cast iron have different thermal expansion rates, so a few degrees difference will affect the measurements.

There are two ways to determine piston-to-bore clearance. The first is mathematical. Measure across the cylinder bores 90° to the wrist pin about 3-in. down the bore to determine maximum diameter. Then, measure the piston that goes in that bore. Mike the piston across its thrust faces, 90° to the wrist-pin axis in the plane of the wrist pin. Subtract this figure from bore diameter to get piston-to-bore clearance.

Directly measuring the clearance is the second method. Install the piston and rod

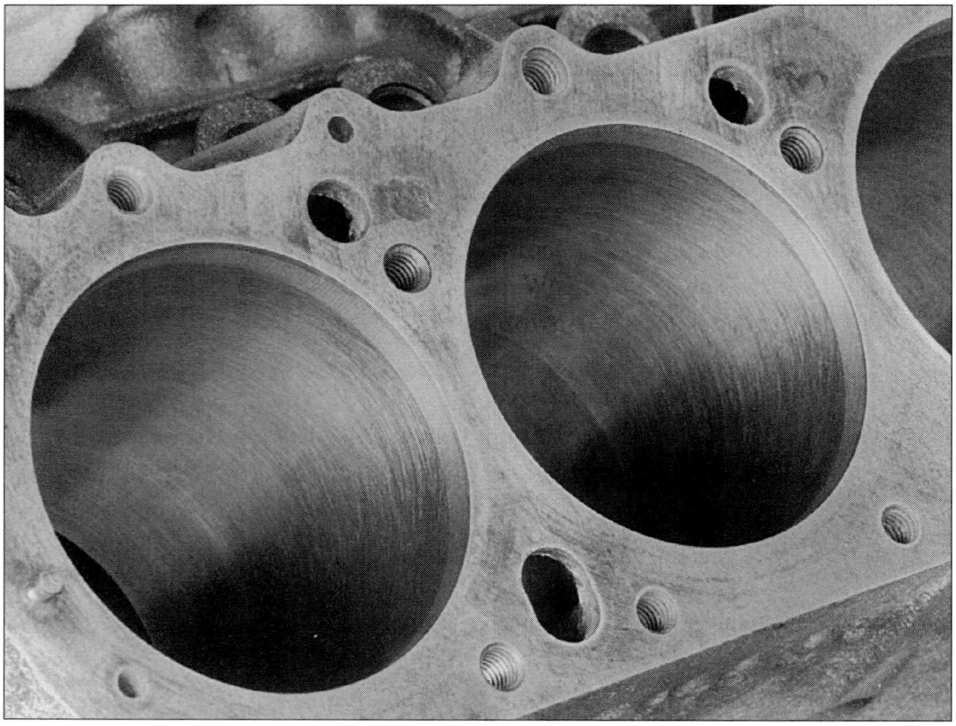

Without honing in this 30° cross-hatch pattern, there would be nothing on the cylinder walls to retain oil. Within a short time, the piston rings would be gone. Here's a trick to remember: the faster you move the hone up and down inside the piston bore, the larger the degree of cross-hatch. Conversely, the slower you move it up and down the smaller the degree.

assembly into its bore without rings and in its normal position. Normal position is with the wrist pin parallel to the crankshaft and with the notch or arrow facing forward.

From the top of the cylinder, slip a feeler gauge between the piston's thrust face and the bore in several positions. Move the piston up and down the bore to check between TDC and BDC. The maximum clearance for either of the methods is 0.008-in.

Knurling—If finances prescribe a super inexpensive job, and piston-to-bore clearance is in excess of 0.008-in., there is one way to avoid buying new pistons. Excess clearance can be taken up with a process called knurling.

In the process, depressions are put in the thrust faces of the piston, creating raised projections that effectively increase piston diameter. The problem is that the projections quickly wear off, and you're right back to square one. So, if you're into the 100,000 mile rebuild, bite the bullet and buy new pistons if maximum clearance is exceeded. You'll be miles ahead!

Honing—If you find no need for boring, the next step is to break the glaze in the cylinder bores. This operation does not change the taper because little material is removed. It does, however, restore a honed finish to the cylinder walls. This serves two purposes. It aids ring break-in and oil retention. Check the cross-hatch pattern at the base of the bores. This is what you'll restore to the cylinder bore when you break the glaze.

A spring-loaded or ball-type hone should be used for this process. Chuck the hone in a variable-speed, half-inch drill and run the hone up and down the cylinder at about half to three-quarter speed of the drill motor. Use a little oil to keep the stone clean.

Continue until there is a good cross-hatch pattern the full length of the bore. If you plan to use plain or chrome rings, this operation is a must. If you're using moly rings, it's not mandatory, but is still good practice.

MACHINING THE BLOCK

There are machine shops everywhere still using equipment and technology left over from a time when machining tolerances were nowhere near as critical as they are today. Although these shops will offer inexpensive work, they may be unable to meet the precision requirements you need to ensure the long life of your engine.

Some machine shops continue to use boring machines that locate on the deck, such as the Van Norman-style boring bar. If the deck is not perfectly square in all directions to the main bearing bores there will be unneeded wear and stress on all of the moving parts. All modern machining locates on the main bearing bores and adjusts everything to this location.

For photographing machine work on the engine block I selected Evans Speed Equipment in South El Monte, California. This shop has been in continuous operation since the post WW II years of the flathead Ford. The shop is now owned by Gene Ohly, who has been turning out premium work since 1963. For over 45 years this shop has produced some of the finest engine machining in Southern California.

FINAL CHECKS

A visual check of the block should reveal any obvious problems, but it is possible to have a nearly invisible crack. The first thing the machine shop will do is to check for these cracks. The process they use is called *Magnaflux.* If you're not going to take your block to the machine shop you can check it yourself with a product called *Spotcheck.* Further more, you'll need to check the deck and lifter bores to determine if these areas need any machining.

Spotcheck seeks the same end as Magnaflux but uses a dye to indicate cracks. Spotcheck consists of a penetrant—a liquid red dye—and a

Following the check for cracks by Magnafluxing, the block is further tested for possible water leaks. The aluminum fixture attached to the block deck surface seals the waterjacket ports. Expansion plugs are inserted in the core plug holes. About 170 lbs. of water pressure is applied to the block through the hose you see at the right. After 15 minutes, if Jeff finds no water leaks, he will pronounce the block fit for machining.

Jeff Elgin, an up-and-coming machinist at Evans Speed Equipment, begins the Magnafluxing operation on the block. This is a big 440 that will be developed as a full-on racing engine with the addition of aftermarket Hemi heads from Stage V Engineering.

developer—a white powder. Cleaner is also included to clean the surface prior to and after the test. Spotcheck can also be used to check rods, heads, pistons and crank. The major advantages of Spotcheck are its ability to check non-magnetic parts such as pistons, and its convenience. Spotcheck is available from Magnaflux Corporation, 7301 West Ainslie, Harwood Heights, IL 60656. (708) 867-8000.

Warp

To check for deck warp, select a machined straight-edge as long as the block deck. Set this on the deck and use a feeler gauge to check for gaps between the straight-edge and the deck.

Measure several different places along the straight-edge, locating it at various angles across the deck. If any gap exceeds 0.007-in. the block will have to be decked or, using the more popular term, milled.

Although the most accurate measurements will be made with a long straight-edge, a 12-in. one will suffice. Instead of the 0.007-in. maximum, use 0.003 in. If warp or notching exceeds 0.015 in., the block should be discarded. Removing more than 0.015 in. could result in insufficient deck clearance—distance between the top of the piston and the block's deck surface.

The 1963-64 Max Wedge block had valve clearance notches machined in the block at the top of the cylinder bores. Whatever is machined from the deck of the block, a corresponding amount must also be machined from these recesses.

Lifter Bores

About the only thing left to check is lifter bore wear. It is highly unlikely these will be worn excessively, as they are so well oiled and lightly loaded. To check them, however, slide a new, unoiled lifter into the bore and try to move it sideways. If there is noticeable movement, the bore is worn and should be honed for oversize lifters. These are available in 0.001, 0.008 and 0.030-in. oversize.

CYLINDER BORE FINISHING

Before you drop the block off at the machine shop to have the cylinders bored, you must first determine which type of piston rings you plan to use. The machinist must hone the bore with a particular finish that matches the type of ring. Hence, a discussion on the different types of piston rings is in order.

Types of Piston Rings

Basically there are three types of piston rings—plain cast iron, chrome or moly. Cast-iron rings break-in quickly, but their life span is short. Chrome rings are the hardest and will last longest in engines that inhale a lot of dust as in off-road vehicles.

The best all-around rings are moly. They require no break-in. A moly, or moly-coated ring carries more of its own oil. As a result, it has a greater life expectancy than a cast-iron or chrome ring. The moly coat has microscopic voids that pick up and carry oil as the piston travels. Cast-iron and chrome rings don't have many voids and depend on the bore's cross-hatching for lubrication.

When the piston travels down the bore on its power stroke, part of the oil is burned away from the cylinder wall. On the return, or exhaust stroke, cast-iron or chrome rings receive less lubrication than the moly ring because the moly ring carries oil.

The lubrication properties of a ring dictate the type of bore finish required. If cast-iron or chrome rings are used, the bore should be finish-honed with a 280-grit stone. A 400-grit stone is used for moly rings. A 30° cross-hatch pattern is used for all ring types.

The machining operation begins with the main bearing caps. The bottoms of the caps are ground until they sit flat on the machine deck. Next, the tops, or mating surfaces, are ground. This will reduce the inside diameter of the main bearing bores, allowing the bores to be machined out to original specs.

Each cap is numbered and sent to the bead-blaster for cleaning.

Main Bearing Caps

Be sure to send the main bearing caps and bolts in with the block—even if you are not having the block align-bored. Remember, all the machining is done relative to the main bearing bores and the machinist will want the caps in place in the proper order to take accurate measurements.

If you're having top-quality machine work done to your block, the machinist will want to do a little work to the main bearing caps before he begins the align bore. For the demonstration block, our super machinist, Dick New, grinds a few thousandths off the mating surface of each cap. This effectively reduces the diameter of the bore. Now, there is sufficient metal to be bored away and still use standard size main bearings. Without this operation the bores would

The caps are torqued into place. Note the addition of 9/16-in. studs for the main bearing caps over the stock 1/2-in. bolts. The block is set into the align-boring machine. Everything is centered, squared and aligned.

After the boring operation is completed, Dick New cleans and polishes the bores with the align-hone.

The main bearing bores are done first so that all further operations can be performed in relation to them. Here, the block is being prepared for machining the cylinder bores. The dial indicator in the center of the photo is being used to square the block with the machine tool, referencing the bearing bores. In less sophisticated operations they would reference the cylinder bores by bolting the boring bar to the block deck surface.

"Up close and personal." The grooves you see in the sides of the valley are there to give clearance to the pushrods going to the Hemi heads discussed earlier.

be enlarged, requiring the use of oversize main bearings.

If you only plan to have the cylinders bored you'll still need to include the main bearing caps. These should be installed and torqued. (383, 400, 413, 440, to 85 ft-lbs.; 426 to 100 ft-lbs.) Torquing the bolts distorts the bores slightly. Simulating these loads during machining results in straighter bores when the engine is assembled.

Some machine shops will also bolt a plate on the block deck to simulate the installed head. This is a good idea on a standard rebuild, but expect to pay for it. Consider it mandatory for high-performance work.

Chamfer the Bores

When you get your block back from the machine shop, check to see if the edges of the bores have been chamfered, or beveled. This chamfer is a 60°, 1/16-in. wide bevel around the top edge of the cylinder bore. It serves two purposes. When installing a piston, it provides a lead-in for the rings. It also eliminates a sharp edge that could glow hot and cause preignition.

If the chamfering has not been done, do it yourself. Use a round or half-round file held at a 60° angle to the deck. File around the entire circumference of the cylinder, but take care not to scratch the cylinder wall opposite the side you're working. Electrical or duct tape over the end of the file will protect the bore. Don't ruin a new bore.

One More Time

Remember my admonishment earlier: there is no such thing as an engine that's too clean! Even though the block was sanitized and wrapped in plastic at the machine shop, it's your engine and you're the final authority on whether it's clean or dirty—so assume it's dirty! During the machining, metal filings find their way into every nook and cranny. These must be removed to your satisfaction. The machinist may have only blown it off with an air gun after he finished the job. So, get out the hot water and soap, brushes and paper towels and clean the block again.

After a thorough cleaning, wipe the block dry with paper towels. Don't use cloth rags; any cloth lint left won't dissolve. If you have it, use compressed air to blow the excess moisture out. When the cleaning is finished, a fresh paper towel wiped through the bore should come out as clean as it went in.

Spray each bore as you finish with WD-40 to prevent rusting, then spray the entire block when it's done. This is very important. Without the oil, rust will develop on freshly machined surfaces within 10-15 minutes after the block is dry. So spray it well. Don't use motor oil, which traps moisture. WD-40 and other water dispersing oils displace the moisture. Wrap the clean block in a heavy-duty plastic trash bag and set it on its rear face until final engine assembly.

CRANKSHAFT

At first glance it might seem that all the big-block cranks might be directly interchangeable, that is until you discover the main bearing journal sizes between the "B" and "RB" block are different: the "B" block crankshaft main bearing journal is 2.625-in. while the "RB" journals are 2.750-in. The rod journals are all the same, however, measuring 2.380-in.

As you would imagine, the crankshaft is the very heart of the engine. Although it takes the greatest punishment, it's rare that a crank wears out or is irreparably damaged. The usual damage is grooved journals from foreign particles that passed between the journal and the bearing.

Replacement is usually required only when another part, usually a connecting rod, breaks and tangles with the crank. A visual inspection of the crank should turn up any obvious damage. Check for cracks by Magnaflux or Spotchecking.

Inspecting Main & Connecting Rod Journals

Begin by inspecting for cracks, surface finish, bearing-journal roundness and taper. A cracked crankshaft can't be repaired. Out-of-round and tapered journals are illustrated on page 79.

Surface Finish—The journals should be smooth with no nicks or scratches. Feel each journal for burrs that could damage the bearings. Oiling-hole edges should be smooth. If they're sharp, a small rat-tail file should be used to chamfer them, much like the cylinder bores.

Check the journal surfaces with your fingers. If you can feel any roughness or imperfections, so will the bearings. If these imperfections cannot be cleaned up with emery cloth, the crank will have to be reground, even if the journals are true. If one journal has to be reground, than all the rod or main journals must be reground. Although this sounds expensive, the biggest cost of the job is setup time. Once the crank is set up in the machine for regrinding, it's a simple task to grind all of the journals. The second reason for grinding all the journals is that the bearing inserts—rods or mains—are sold in sets. There are few parts houses that will break a set of bearings and sell only one.

Grinding off 0.010-in. will clear up most crank-journal problems and is pretty much standard practice. If the journals are ground 0.010-in., you must order 0.010-in. undersize bearings. Undersize refers to the journal diameter, not bearing thickness. If, for instance, your 440 crank must have the main bearing journals ground, then 0.010-in. from 2.750-in. leaves undersize main bearing journals of 2.740-in.

Out-Of-Round—When you cut an orange in half and just begin to squeeze out the juice, the oval shape of the orange is an ellipse. In a cutaway view of a journal, if this elliptical condition exists, the journal is out-of-round. Maximum allowable out-of-round is 0.001-in. with the exception of the 426 Hemi. Maximum out-of-round for it is 0.0003-in.

Taper—Taper is another problem encountered with the bearing journals. In this case, the diameter of the journal varies from one end to the other. If you measure a journal and there is a difference in readings between one end and the other, the journal is tapered. Like out-of-round, maximum taper is 0.001-in.—and in the case of the 426 Hemi, 0.0005-in. is the limit.

To surface the deck, the block is again squared to the machine in reference to the main bearing bores. Then it's leveled front-to-rear and side-to-side. A number of passes will be made taking off one or two thousandths at a time until the deck is perfectly flat.

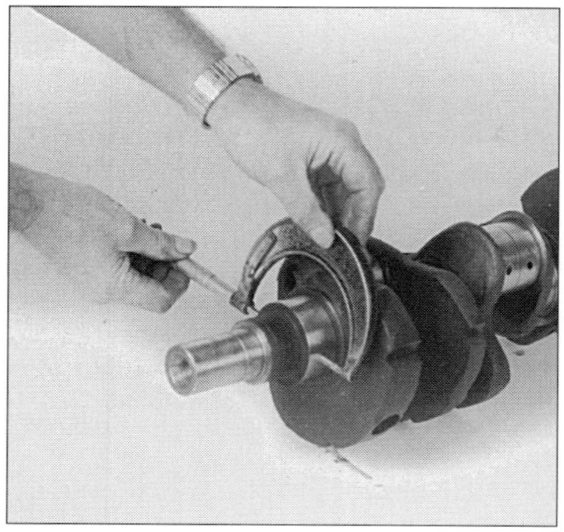

You must have a good micrometer to check the crank for taper and out-of-round. As you take the measurements, be sure to write them down for later reference.

Polishing journals is really a lot of fun. You can do it in time to your favorite song or whatever is on the radio. The faster the music the better the job! Be sure to use an emery cloth no courser than 400-grit.

Note the Mallory metal (arrow) insert in the crank throw. Along with metal removal (drilled holes) the two work together to fully balance the crankshaft.

The crankshaft for this big sucker is being reconditioned at Velasco's Billet Crankshafts in Downey, California. Henry Velasco begins by centering the crank in the grinding machine using a dial indicator. In the case of this crank, it will be set off-center. When ground, it will change the center of the journal, thereby lengthening the throw and increasing compression. Your crank will be ground the same way only fully centered. By the way, the reason the crank looks so pretty is that it was bead-blasted first.

To check both out-of-round and taper, measure a journal in several places. Use a 2 to 3-in. micrometer. Measure each end of the journal and in the center. Measure in several places around the diameter. Check for both out-of-round and taper.

As you measure around the journal, you will get different readings—write them down. After taking several readings all around the journal, subtract the smallest reading from the largest one. This difference is the amount of out-of-round. Taper is the difference in size from one end of the journal to the other.

Suppose after measuring a rod journal on a 383 crank, you get readings of 2.373-in., 2.371-in., 2.374-in. and 2.375-in. By subtracting the smallest, 2.371 from the largest, 2.375, you find the rod journal to be 0.004-in. out-of-round or tapered. Chrysler specifications for out-of-round and taper give a maximum of 0.001-in. for both rods and mains. As a result of these readings, you would have to regrind the journals.

If you determine your crank must be reground, take it to the best crank shop in town. You'll be far ahead. If saving money is your main concern, then trade your old crank in for a crank kit. This is a reground crank with the rod and main bearings included.

Thrust-Face Width—Pay particular attention to the number 3 main bearing journal if your engine is backed up with a manual transmission. When the clutch is disengaged, the crankshaft is pushed forward. This load is taken by the crankshaft's rear thrust surface, at the number-3 main bearing.

If the pressure plate had an excessively high load or the clutch was misadjusted, the thrust surface may be damaged. One indication of possible thrust-surface

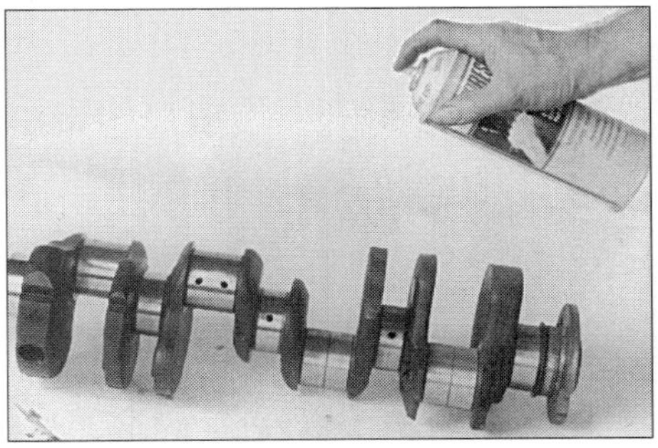

After polishing the journals you must use a water repellant to prevent rust. When finished, wrap it in plastic and stand it on its flange.

The grinding machine in action. The fixture riding on the bed of the machine helps hold the crank journal in perfect alignment against the pressure of the grinding wheel.

With this machine Chris can polish all of the journals faster than you can do one by hand. He, too, uses emery belts under 400-grit.

trouble is excess wear on the pressure-plate release levers.

Minimum width between the thrust faces is 1.152, maximum is 1.158. Check the thrust-face widths in several places. If the 1.158 figure is exceeded, you'll have to replace the crank or have it reconditioned.

You'll need a 1-2 in. inside micrometer or a 1-2 in. outside mike and a telescoping gauge or a bearing insert and feeler gauges. With an inside mike the thrust surfaces can be measured directly. With a telescoping gauge, simply measure between the thrust faces, then check the gauge with the outside mike.

If you're using a bearing insert and feeler gauges, mike the insert across the thrust surfaces and record the width. Place the insert on the journal and use the feeler gauges to check the distance between the insert and thrust surface. Add this to the insert width to get thrust surface width.

Check Finished Journals— Whether or not you've purchased a new or reground crank the journals may not be as smooth as they should be. The easiest way to check this is to draw a penny across the journal. If any of the copper from the penny rubs off, the journal is too rough. More than likely, you can fix it yourself. All you'll need is about 3-ft. of 1-in. wide, 400-grit emery cloth.

Set the crank on a clean bench. Loop the emery cloth around the journal, take up the slack, and begin zipping it back and forth as you see me doing in the photo on page 76. Work your way all around the journal. The object is a smooth finish with no nicks or burrs.

Crankshaft Run-Out

Crankshaft run-out is the amount the crankshaft is bent. There are two ways to check run-out—the dial-indicator method or installation method

Dial-Indicator Method— Install only the old front and rear bearing inserts in the block, oil them and place the crank into the block. Mount a dial indicator on the block. Place the plunger off-center on the number 3 main bearing journal to miss the oil hole. Rotate the crank until

Henry mics the journal. He wants to leave 0.0015-in. over so there will be material left to polish away with a finer grinding wheel.

Check run-out with a dial gauge. Use only the front and rear main bearings to hold the crank. As the crank is rotated, run-out should not exceed 0.004-in.

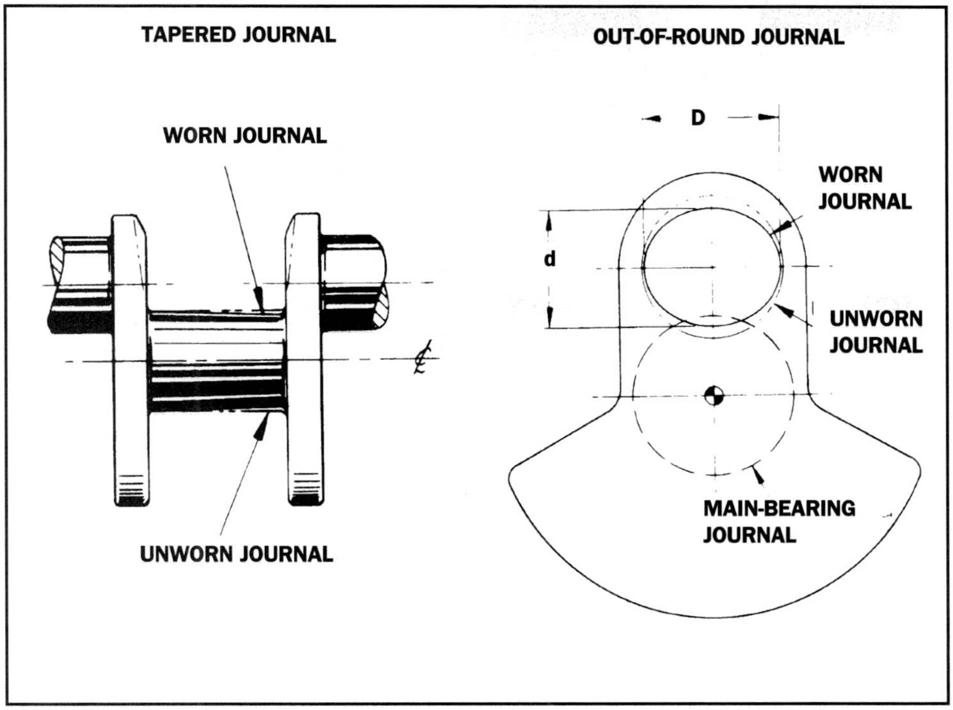

Tapered journals wear across their diameter. Out-of-round journals become elliptical or oval shaped. Connecting rod journals tend to wear out-of-round due to the way they are loaded.

you find the lowest reading. Now, zero the dial and rotate the crank again. The maximum reading is the amount of run-out.

Maximum allowable run-out is 0.004-in. If the run-out exceeds this amount, don't panic yet. Turn the side with the maximum run-out up—away from the block. Install the center main cap with the bearing in the cap. Torque it to 85-ft-lbs. and leave it a day or two. This should pull the crank into position. Recheck the run-out. It should be back within the 0.004-in. limit. If it's not, have it ground or trade it for a crank kit.

Installation Method—The second and more practical check is to install the crank without the oil seal but with new, well-oiled bearings. Torque the caps to spec. If the crank rotates easily, you're home-free. Any run-out drag is minor compared to the loads produced under power. If you decide to use this check, follow the crank installation procedure on page 120.

Check After Regrinding

As much as we'd all like to be perfect, we can't be. Neither can the machinist. When your crank comes back from the shop you must check it out. There are two areas you must check: the edges of the oil holes and the shoulder radii at the ends of the journals. Sharp edges on the oil holes will score a bearing as fast as a metal filing between the journal and bearing.

Wrong shoulder radii will cause different problems. If the radius is too narrow or nonexistent, it will weaken the crank at this point. This could result in a crack and ultimately a broken crank. If the radii is too large the crank will do just fine but it will edge-ride the bearing—the journal will ride on the edge of the bearing—wearing both bearing and journal.

Maximum main and rod journal radius is 0.100-in. If you find any that are not right, take the crank back.

More Cleaning—Here we go again! You must continue to clean all the parts. For the crank, the oil holes are the important areas. Soak a rag in carburetor cleaner or lacquer thinner and pull it through each hole with a thin wire until it comes out clean. If you have a copper bristle brush, use it. It'll do a great job. Clean the outside of the crank with lacquer thinner and finish off with a good coat of water dispersing oil.

Pilot Bushing

If your car is equipped with a manual transmission, it most likely has a crank with a pilot bushing in the center. I say most likely, because some of the later model 440's in family station wagons omitted this.

The pilot bushing supports the forward end of the transmission input shaft. It's an important piece and should be replaced.

There are three ways to remove the pilot bushing: with a slide hammer, bolt or chisel. A slide hammer is a shaft with a hook on one end, a handle on the other and a weight that slides between them. Insert the hook through the bearing and catch the back edge. Then, slide the weight back against the stop on the shaft. The impact will pull the bearing free. Of course, it will take a few smacks to do it.

The bolt method requires a course thread bolt that can be screwed into the bushing. When it bottoms out against the crank, continue turning the bolt until the bolt threads pull it out.

I like the third method—splitting it. Use a small chisel (a cape chisel is best) to split the bearing. Be careful not to scratch the bore. When you've cut two grooves, pry the two halves out with a screwdriver.

The new bushing should be soaked in oil for about 15 minutes before you install it. The Oilite, or sintered-bronze bushing, is self-lubricating, but any help it can get will be greatly appreciated.

Grease the bore and start the bushing in with a plastic mallet. It's very soft so use care. Now, with your bushing driver and a 12 to 16-oz. hammer, drive it home.

CAMSHAFT & LIFTERS

Well, we're on the downhill side of inspection and reconditioning with the completion of the crankshaft. The camshaft and lifters will be a piece of cake by comparison. There are two rules you must remember, however, to insure success.

1. When using an old camshaft and its lifters, never install the lifters out of order.

2. Never install used lifters on a new camshaft.

Following these rules will reduce the chance of cam-and-lifter destruction on initial startup. Normal operating loads are extremely high. Poor installation guarantees you'll ruin the cam during the first 30 minutes of engine operation. Follow instructions in the next paragraphs religiously.

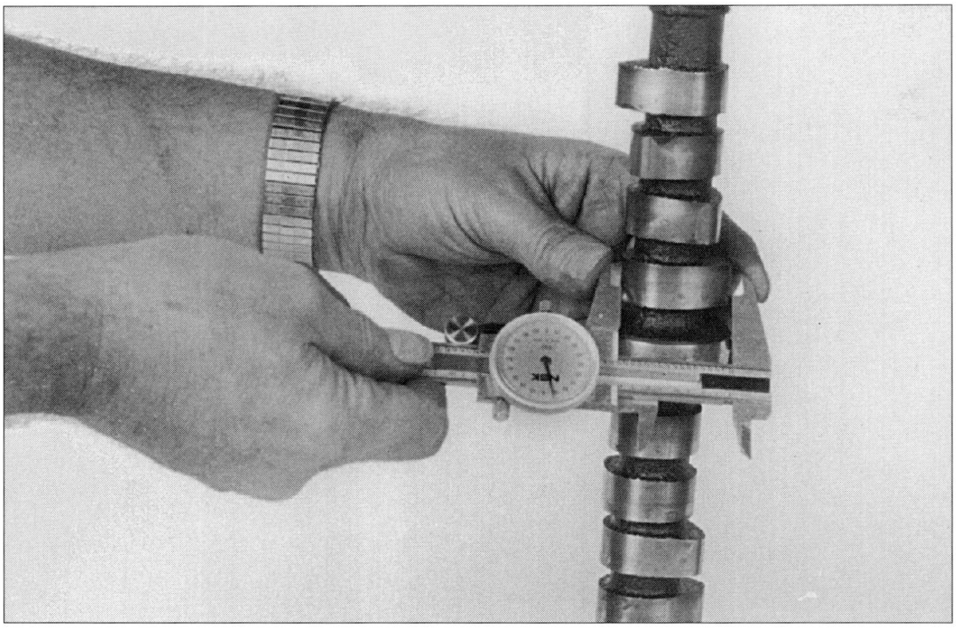

I'm attempting to determine camshaft lobe lift by measuring each lobe at its smallest and largest diameters. Difference should not exceed 0.005-in. from one lobe to the next. Be sure you compare intake with intake and exhaust with exhaust.

Design

A quick course in cam and lifter design will help you understand why these areas are so critical. Follow the discussion below with the drawing on this page.

Notice the cam-lobe profile is egg-shaped as viewed from either end of the cam. As the cam lobe rotates, the opening ramp raises the lifter in its bore, opening the valve through the pushrod and rocker arm. The lifter and valve reach maximum travel when the lifter is on the toe or nose of the lobe. The valve returns to its closed position as the cam continues to rotate and the lifter rides down the closing ramp to the base circle or heel of the lobe.

Changing the shape of the lobe changes the rate at which the valve opens. It also changes the length of time—duration—the valve is open or closed. Not only can the duration be changed, but the distance the valve opens—valve lift—can also be changed. Changing duration or lift affects the amount of power and where it occurs in the rpm range.

When the valve is closed, the lifter is on the base circle or heel of the lobe. It rides up the ramp to the toe and opens the valve through the lifter, pushrod and rocker. The cam lobe determines the cam's lift at the lifter, not at the valve. Valve lift depends on lobe lift and the mechanical advantage—usually expressed in a ratio, like 1:1 or 1.5:1—of the rocker arm. (Using higher ratio rocker arms such as 1.6, 1.65 and even 1.7:1 increases valve lift.)

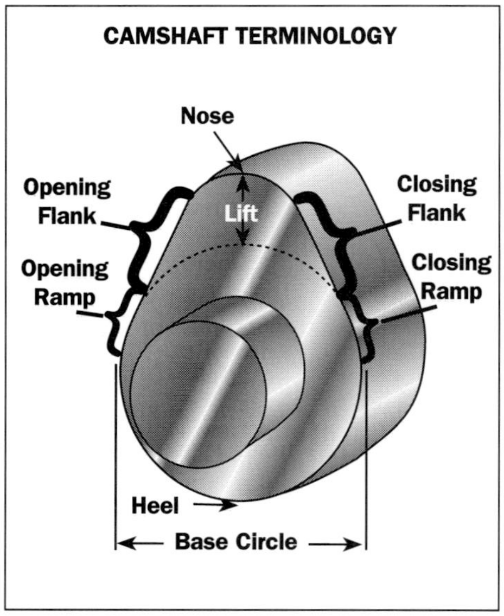

This drawing illustrates the terms used in this section. Any measurable wear you find will mean the cam and lifters must be replaced—and always as a group.

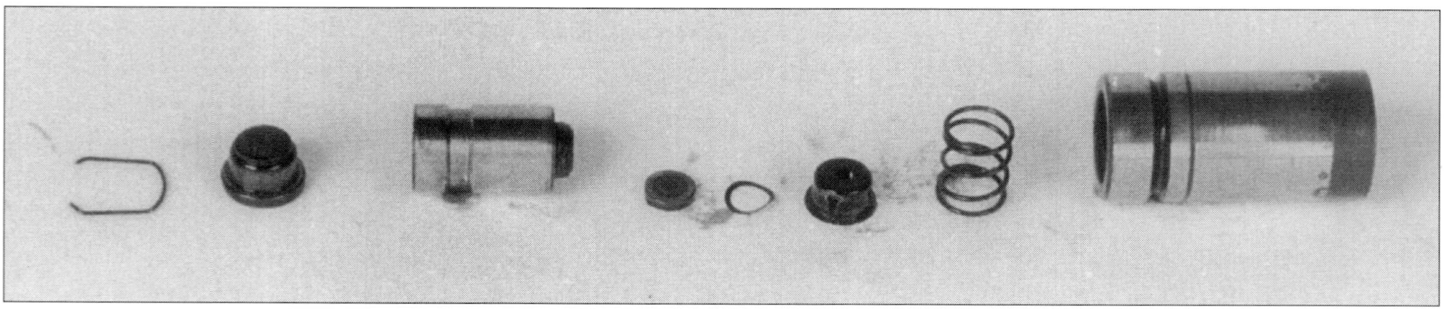

I opened this lifter and laid the parts out in order. Now, if you drop a piece on the floor you can see where it goes.

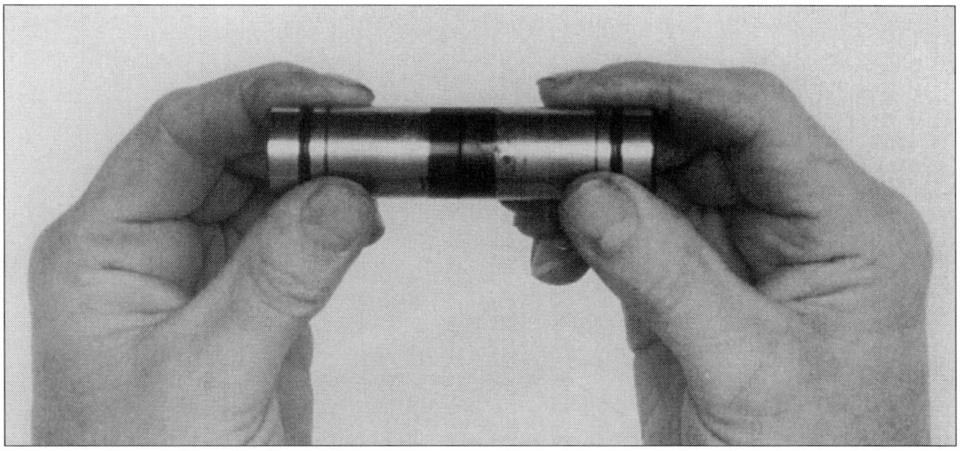

To determine if the lifter foot has been worn flat, butt the two ends that ride on the cam lobes together. If they rock around a bit they're still in good shape. If there's no movement at all, they're scrap metal.

A cam lobe surface is not ground parallel to the cam centerline, but at a rake angle—usually 1°. The lifter base or foot that contacts the lobe has a convex spherical radius, like a magnifying glass.

Lobe rake angle and lifter-foot radius ensure tolerable contact pressure with good lubrication. Secondly, the lifter is placed slightly off-center to the lobe. This spins the lifter as the cam rotates, which increases cam and lifter life.

Inspection

Here's some good advice: toss your lifters, buy a reground cam and don't waste your time reading this section! Wow, pretty straightforward. Yes, straightforward because the cam is worn. How much wear depends on engine use and how well it was maintained. Rule number 2 says never install used lifters on a new camshaft so both should be replaced. Okay, so you don't want to take my advice and want to know how to check for yourself. Let's start with the cam.

Checking the Cam—Check for visible wear and pitting. If the toe of any lobe is worn completely across the face the cam is junk. Any pitting on the lobes means metal loss and, again, the cam is worthless.

The next thing to look for is measurable wear. If you used a dial indicator to check the cam as outlined in the diagnosis section, you know where you stand. If not, get out the mikes or vernier caliper and check the lobes to compare lifts.

Measure the base circle diameter across the lobe—90° to the line through its toe and centerline. This is the lobe's minor dimension. Next, measure the major dimension on the line through the toe and cam centerline.

Make both of these measurements the same distance from the edges of the lobe where there is maximum visible wear. Subtract the minor dimension from the major dimension to get approximate lobe lift.

Note: this is very inaccurate when dealing with high-lift, high-performance camshafts, because the ramp extends farther into the base circle. It can, however, be used to compare lobe wear.

Measure each intake lobe and write down its lift. See the chart on page 10, Chapter 1 for specified valve lift. Do the same with each exhaust lobe. To make it easier for you the lobes are EEIIIIEEEEIIIIEEE from either end. Now, compare all intake lobe lifts and all the exhaust lobe lifts. Be sure you are comparing intake with intake and exhaust with exhaust. Lift should be within 0.005-in.

If any of the lobes show more than 0.005-in. wear, trade it for a reground or new cam.

Cam Bearing Journals—Cam bearing journals are basically indestructible. You can, however, check them for wear if you wish. Use a 1-2-in. outside micrometer.

There are five bearing journals each of a slightly different size. The following chart gives the cam journal diameter limits for all "B" and "RB" engines. From front to rear they are:

1. 1.998 to 1.999-in.
2. 1.982 to 1.983-in.
3. 1.967 to 1.968-in.
4. 1.951 to 1.952-in.
5. 1.750 to 1.751-in.

Run-Out—The best check for run-out is to install the cam with well-oiled bearings and check that it spins freely. If you wish to check for journal out-of-round, the limit is 0.005-in.

Lifters

Aside from obvious damage, scoring, stuck plungers or broken springs on the hydraulic lifters, check the foot radius. If it's gone, all you have is junk.

Check the radius by placing two lifters foot-to-foot and try to rock them back and forth. If there is noticeable movement, there is some radius left. If not, they are flat or even concave. While you can run new lifters on an old cam, a worn lifter foot will also wear the corresponding cam lobe. I hate to keep preaching this gospel but if one lifter foot is shot, the whole works should be replaced: one cam and sixteen lifters.

Clean the Lifters—If you haven't thrown away the cam and lifters thus far then you can turn your attention to cleaning the lifters. Continuous cleaning and checking are the trademarks or a conscientious mechanic; the result is another 100,000 miles on the new engine.

Solid lifters can be cleaned with carburetor cleaner or lacquer thinner. Hydraulic lifters should be disassembled and thoroughly cleaned.

With a pair of needle-nose pliers, remove the retaining clip and slide the lifter apart. The photo on page 80 will help you get it all back together. If the lifter refuses to come apart and a gentle rap won't persuade it, soak it in lacquer thinner. This should free it up. Wash the parts thoroughly. To keep from mixing lifter components, clean one lifter at a time. Oil the parts to prevent rust and aid reassembly. Although you must repeat this 16 times, you'll rest easier as you drive, knowing those little rascals are doing what Chrysler meant them to!

PISTONS

If the block does not require reboring, the next step is to check the pistons. Keep the pistons and rods intact for now but remove the rings. Be careful not to scratch the pistons.

The easiest way to remove the rings is to clamp the rod in a vise—not too tight now—or to the side of the bench with a C-clamp. Rest the piston skirt against the tip of the vise or bench so the piston won't rock. Slip the rings over the top of the piston. Sliding them down over the skirt may scratch it and ruin the piston. A ring expander will make the job nearly foolproof. Ring edges are sharp, so if you do it by hand you'd better have some good calluses or a high pain threshold. Catch the ends of the ring and spread it as much as possible. Push it back out of the groove and rotate it over the top or dome of the piston.

Any of four basic problems require piston replacement: damage to the dome, skirt or ring lands; ring-groove wear; piston-skirt wear or pin-bore wear.

Piston Damage

Obvious damage—a valve head implanted in the dome or a chunk of skirt missing—means piston replacement. Not so obvious is skirt scuffing or scoring, skirt collapse, ring-land damage and dome burning. (Ring lands are the spaces between the ring grooves—the "land" between the grooves.)

Scuffing and scoring—vertical scratches or marks—are readily visible even to the untrained eye and require piston replacement. Scuffing and scoring are usually caused by one or more problems: lack of lubrication, excessively high temperatures or a bent connecting rod. Any of these cause excessive friction between the piston and bore.

If the scoring or scuffing pattern is asymmetrical around the skirt thrust faces, check for a twisted or bent connecting rod. It's possible to have a bent rod, even a drastically bent rod straightened. Again, this is your call based on how the engine will be used and how long you want the engine to last. The cost of a new rod is not much more than the labor and machining cost to straighten your old one. Whatever you decide, be sure you have all the rods checked by your machinist.

Clean the top of the piston by scraping away the carbon deposits and wire-brush it to remove any leftovers. This will provide a good view of the dome and any problems.

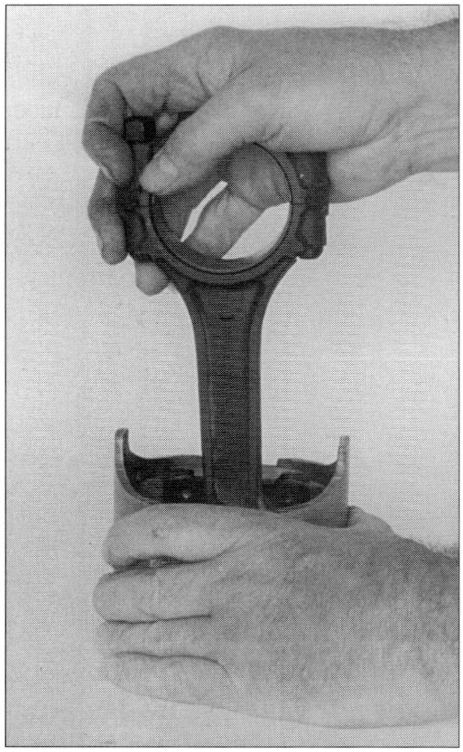

Before you make this test get all of the oil out between the pin bore and pin. Do this by soaking the piston in solvent for a little while. Now, dry it out, set it on the table and try to wiggle the rod in all directions while holding the piston flat on the table. Any movement means the bore or pin is worn.

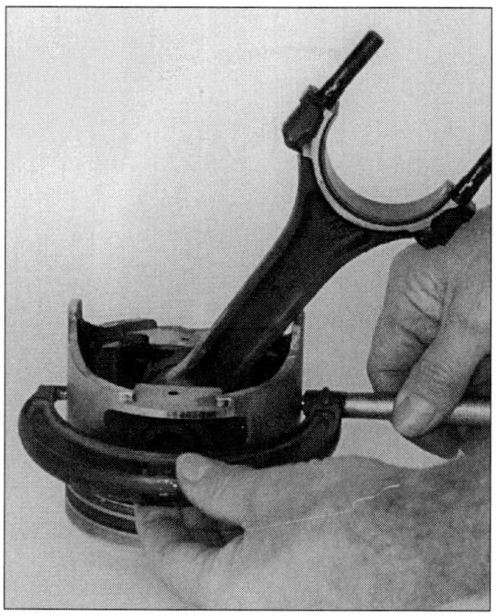

To determine whether or not the skirt on the piston has collapsed, measure it first 90° to the pin and then again at the bottom. The skirt should be at least 0.0005-in. larger at the bottom than at the pin. Four places is pretty hard to catch. Be sure there is no dirt on your piston or micrometer.

It is said that the intelligent person learns by another's example. If you're smart, you'll learn from this example. I got careless and accidentally dropped this piston, breaking the skirt. Now it's worthless. I hope you'll be more careful!

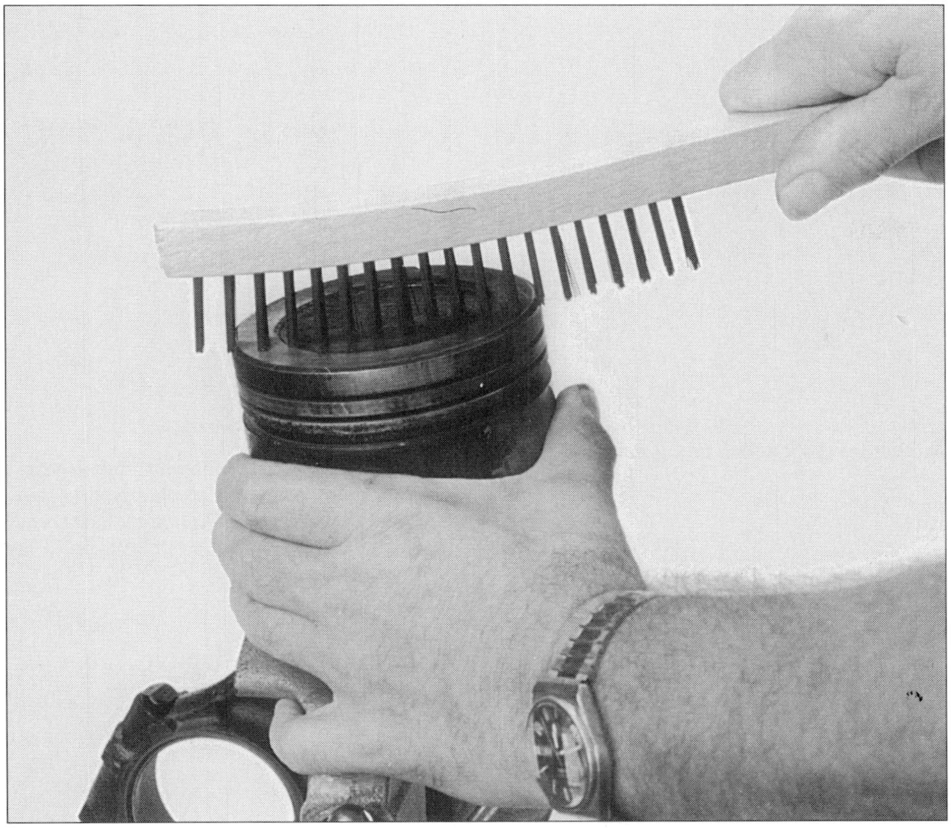

Yes, it is a new wire brush; the other was just too far gone from cleaning the barbecue. Nevertheless, if you plan to use your old pistons over they must be thoroughly cleaned. A new wire brush is a small price to pay.

You never know what kind of damage will happen to a piston. I was told by the mechanic that removed this piston that the bolt shown imbedded here fell down through the carburetor. It sure did a number on the top of the piston. The cylinder bore doesn't look much better.

Lean mixtures, retarded timing or preignition will often cause the dome of the piston to overheat and burn. This is indicated by rounded edges on the dome or spongy-looking material where the heat was concentrated.

Preignition or detonation—pinging—can break or distort ring lands through impact loading. Usually a broken land is readily visible. A bent land, however is difficult to see without a ring in place.

To check for warped or bent lands, use the checking procedure that follows.

Piston Skirt Diameter

Drag out the micrometer again and measure the piston skirts. A piston is not a true cylinder; it is flared—wider—at the bottom of the skirt.

Mike each piston 90° to its wrist-pin axis 1/4-in. below the oil ring land. Compare this figure to the diameter across the bottom of the skirt. If the skirt is not at least 0.0005-in. larger, it is partially collapsed and the piston should be replaced.

Skirt collapse is usually indicated by heavy scoring or scuffing on the skirt—sure signs the engine was severely overheated at least once. When an engine overheats the piston expands until the skirt contacts the bore, squeezing out the oil film between the skirt and bore. This often results in scuffing, scoring and possible skirt collapse.

When the engine cools, the piston retracts to a smaller size and the excess skirt clearance causes piston slap. Reinstalling this piston would result in a very noisy engine. The skirt would eventually break off.

Piston Pin Bore Wear

The 383, 400, 413 and 440 pistons use pressed in or "interference fit" wrist pins. The 426 is the only piston in the "B/RB" series that uses a "free-floating" pin. Interference fit pins must be removed with a hydraulic press while the free-floating pin can be extracted by first removing the snap ring retaining it inside the piston bore. Once out, clearance can be measured.

Pin-to-bore clearance should be 0.00045-0.00075-in. on pistons with the pressed-in pins—the interference fit is between the pin and the rod. Clearance on the free-floating pin of the 426 is 0.0001-0.0006-in.

For "rule-of-thumb" measurement, wash the piston and rod in solvent to remove the oil between the pin and piston. Oil takes up clearance and cushions the movement. Dry the piston and place the dome down on a flat surface.

Hold the piston firmly with one hand, with your thumb on one end of the pin and a finger on the other. With the other hand try to twist the rod 90° to the pin, then wiggle it back and forth. If no movement can be felt, the pin-to-bore clearance should be okay.

Checking & Cleaning Ring Grooves

Now, assuming that the pistons survived the first part of the inspection, move on to step two—checking the ring grooves. The ring grooves must be cleaned thoroughly because the rings must seat properly in the grooves to seal

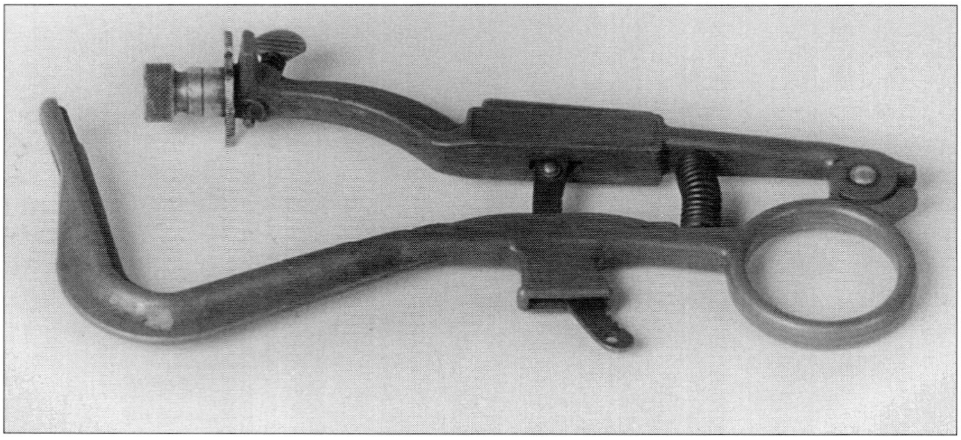

This terrible-looking surgical instrument is really nothing but a high-tech ring groove cleaner. A broken ring really works just as well.

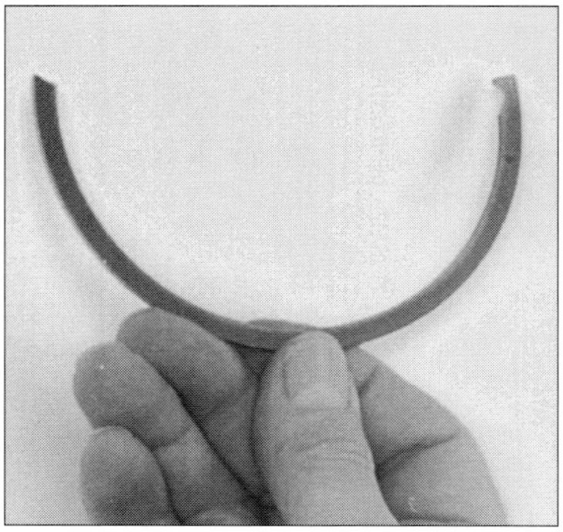

This is the best (and cheapest) ring groove cleaner there is. Break an old compression ring in half and grind a little hook in the end. That real fancy one to the left belongs to my brother.

the piston to the bore.

A special tool called a ring groove cleaner can be purchased at the parts store or rented at the equipment rental company. This tool adjusts to fit all standard-size grooves. It's designed to track in the groove on one side and scrape away the carbon and oil sludge on the other.

If you're in a hurry, or don't wish to lay out the cash, you can make a satisfactory ring groove cleaner from a broken ring. Using it requires a little more care and patience than the groove cleaner, but it will do the job.

Whichever method you decide to use, take care not to remove any metal. Each ring must seal against the lower surface of its groove, and any nicks in this area will break the seal.

Measuring Ring Groove Width— If there's ring groove damage, it will probably be at the top groove because of the higher loads here. This doesn't mean you can skip checking the bottom two. Anything can happen, so check each groove carefully.

Next you will be checking ring side clearance—the distance between the ring and the bottom or top of its groove. Too much wear in this area causes improper ring-to-piston sealing. Also, the excess play will shorten the life of the ring and piston, increasing the possibility of ring or land breakage.

Compression Ring Grooves— Chrysler calls for a new compression ring width of 0.0775-0.0780-in. with a maximum side clearance of 0.004-in., except for the 426 which is 0.001-0.003-in. Maximum width for compression ring grooves is 0.080-in.

If you invest in new rings now, then find several worn pistons, you'll be out a new set of rings. So you'll want to use an old ring to check groove width. But how can you measure ring side clearance or groove with a worn ring?

The solution is not difficult—measure the old ring. Suppose the old ring is 0.0768-in. thick. Subtract this figure from the new ring standard of 0.0775-in. This leaves 0.0007-in. of ring wear. If you add this to the side clearance for a new ring—0.004-in.—you have a side-clearance measurement of 0.0047-in.

Four digits is getting a bit nit-picky, so you can round this off to 0.005-in. We'll call this figure *checking clearance* to describe its function more accurately. Let's put all of this into a formula. Checking clearance (CC) = 0.004 + 0.0775 minus the checking-ring width.

Here's the way to measure ring-groove width with the feeler-gauge-and-ring method. Be sure the feeler gauge is below the ring.

Here it is as a formula:

Checking Clearance = 0.004 + (0.0775 – 0.0768)
CC = 0.004 + 0.0007
CC = 0.0047
CC = 0.005 rounded off

Enough of the math lesson, let's check side clearance. Insert the old ring into the groove. It's not necessary to install the ring, simply slide the edge into the groove and work from there.

Staying with the aforementioned example, insert the 0.005-in. feeler gauge between the ring and the lower side of the ring groove. Slide the gauge and ring completely around the groove to check for any clearance variations and to assure yourself the land is free from distortion or uneven wear. The gauge-and-ring combination should slide around the full circumference without binding. Be sure to check both compression grooves.

Oil Ring Grooves—The operation here is quite similar to checking compression ring side clearance. Oil rings are well-lubricated and are not as heavily loaded as the compression rings, so there is very little wear. A visual inspection should be sufficient.

To double-check, lay two old compression rings together and mike their combined thickness. Suppose they measure 0.155-in. Subtract this figure from 0.191-in. to get feeler gauge thickness of 0.036-in. Check the same way as compression ring grooves. If the oil-ring grooves do not exceed 0.191-in., you can be certain of a good fit.

Groove Inserts—Should you wind up in the unlikely situation of having everything within specification except for ring groove wear, you can have ring inserts installed. Piston ring grooves are machined wider and the inserts—usually 0.060-in.—are installed to take up the additional space.

Like the other makeshift approaches I've mentioned, you get what you pay for. The cost of honing the block, machining the pistons, and installing inserts is about half the cost of a rebore, new pistons and rings. You can also expect about half the service from a ring insert job compared to a complete rebore.

If you checked the block and discovered it needed a rebore, obviously the pistons must be replaced. If the block checked out okay, but you find the pistons must be replaced, have the cylinders bored to remove any taper. The big expense is new pistons, so give the rings a chance with straight bores and oversize pistons.

Disassembly & Reassembly

Unless you're rebuilding a 426 Hemi, your pistons have pressed-in wrist pins and disassembly and assembly will have to be performed in the machine shop with a hydraulic press. The pins will have to be pressed out and pressed back in. Never, never attempt to drive the pin out. Do so and you'll have eight expensive paperweights.

Assembling a rod and piston with a floating pin (426) is no problem. Simply reverse the disassembly process. Just be sure the retaining rings are seated in their grooves. If you're assembling your own pistons and rods for your 426, note each piston's relative position to the rod as it is installed in the engine. Each piston has a notch or arrow that must point to the front of the engine. Likewise, each rod is numbered and this number must point away from the camshaft. Follow the demonstration on page 86.

CONNECTING RODS

There are three major areas of concern when inspecting connecting rods. These are out-of-round or enlarged bearing bores, twisting or bending and cracked rod bolts or bolt holes. These may be inspected by eye but you'll be far more accurate using Spotcheck or having them Magnafluxed.

A bent rod will be indicated by piston and bearing wear patterns. This is why you should save the bearings. A bent rod creates side loads that cause uneven wear on the connecting rod bearings. Piston skirt pattern will be skewed.

A tapered crankshaft bearing journal will give a similar bearing-wear pattern, but the wear will be on the same side—front or rear—of both inserts. A bent rod will wear opposite sides of the inserts.

As an example, if a rod is bent, the upper insert will be worn on the rear portion while the lower insert will be worn on the front. Double-check against the crank. If you suspect a rod is bent or twisted, either replace the rod or as a second choice, take it to the machine shop and have it checked and straightened.

I suggest you take all the pistons and rods in for reconditioning as a matter of course if you're going for the 100,000 mile job. During reconditioning, the machine shop checks the connecting rod bearing bore for out-of-round or oversize. An out-of-round—oval or egg-shaped—bore wears the bearings unevenly and accelerates bearing wear.

All but the 426 Hemi have interference-fit wrist pins. They must be removed with a hydraulic press. Never, never attempt to drive them out with a hammer and punch!

PISTON & ROD ASSEMBLY

FREE-FLOATING WRIST PINS (426 HEMI)

Remember when assembling pistons and rods that each rod cap has a V-groove cut into the parting face on one side. When the cap is installed on the rod, this groove will point inboard. The groove will point to the opposite row of pistons. The cap number will point left for the left side and right for the right side—and be lined up with the rod number.

With all the parts arranged in the position in which they will be installed, you can now begin putting them together. Install one retaining ring in each piston. Start with the left bank. Keep the large chamfer in the same direction as the piston arrow or notch, insert the rod into number-1 piston.

Thoroughly lubricate a wrist pin and slide it into the piston and through the small end of the rod. Install the second retaining ring to keep the wrist pin in place.

Proceed with the other three assemblies for the left bank in the same way, then do the right bank.

If these procedures are followed carefully each piston will have its notch or arrow pointing to the front of the engine. Each rod will ride against its respective radius on the crankshaft, and each rod will oil the piston immediately to its right or left.

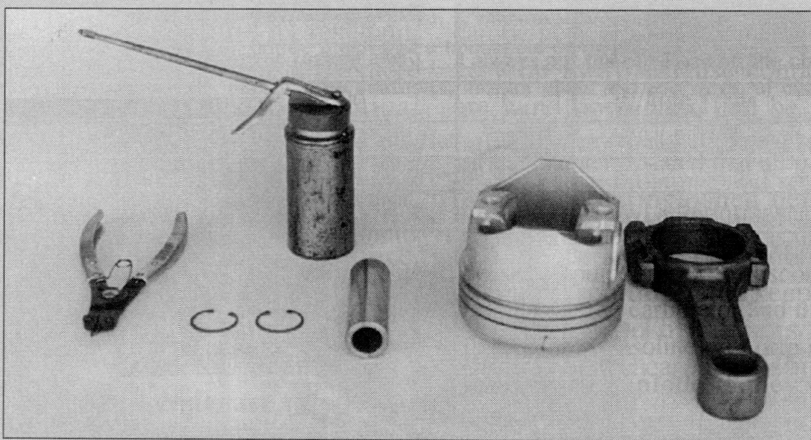

1. We're ready to assemble the rod to the piston. Be sure you have some oil, snap rings and snap-ring pliers.

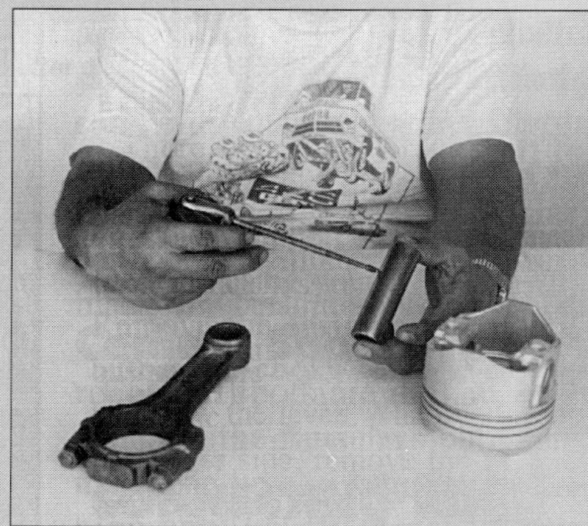

2. It's a long time before oil gets to the wrist pin when you first start the engine. Therefore, be sure you oil all mating surfaces fully before you begin assembly.

3. After placing a retaining ring in one side of the piston, slide the oiled pin into the opposite side.

4. Place the rod into the piston, per the above instructions, and slide the pin through.

5. Finish the job by placing the remaining snap ring (retainer) in place. Check to be sure everything is correctly aligned.

The too-large bore is worse. Bearing inserts are manufactured slightly larger in circumference than the bearing bore so they will force together (crush) when the bearing cap is installed and torqued. Bearing crush locks the inserts in place and prevents them from moving.

If the big-end bore diameter is too large—more than 0.001-in. oversize—the bearing halves will not crush. The bearings will then actually move, leaving shiny spots on the back side of the bearings. If not corrected the bearing will knock or spin—rotate in the bore. If the bearing spins, the crank and rod may be destroyed. An oversize bore is corrected during reconditioning.

Chrysler uses a locking tab at one end of each insert to locate and help prevent them from spinning, but this won't keep the bearings from spinning if the bore is oversized.

When a connecting rod is reconditioned, some of the material is precision-ground from the bearing cap mating surfaces. The cap is then reinstalled, the bolts torqued and the bore honed to the correct diameter and concentricity.

Rod Bolts

Rod bolts are a frequently overlooked item. A rod bolt breaking at any rpm can wipe out an engine!

Obviously you must remove the rod bolts to inspect them. However, once the rod bolt has been removed, the rods must be reconditioned. Reinstalling the bolts will shift the bearing cap slightly and will require reconditioning the rod bores to ensure concentricity.

Remove the rod bolts and look at them carefully through a magnifying glass. If you want to be real careful, use Spotcheck on them or have them Magnafluxed along with everything else. While you're looking for cracks, inspect the surfaces around the bolt holes. Cracks here are rare, but it won't hurt to check.

To remove the bolts, clamp the big end of the rod in a vise between two wood blocks to cushion it. Tap the bolt out with a soft hammer—copper or lead—or a wood, leather, or plastic mallet. They'll be pulled back in when they are torqued. If you find a suspect bolt, replace it.

Clean and oil the bolt threads and torque the nuts 40-45 ft-lbs. before taking the rods to the machine shop. Be sure each cap is correctly mated to its rod.

To recondition the rods, the connecting rod bearing bore and the wrist-pin bore must be honed to their specified diameters and concentricity. Usually the wrist-pin bore will only need polishing and any burrs removed.

OIL PUMP & DRIVESHAFT

The oil pump in the engine is like the human heart in the body. Both pump vital fluids to all the parts. When either pump quits, death is the outcome. In the case of our engine, oil stops flowing, heat builds up and the engine seizes. I think we all have a story about losing oil

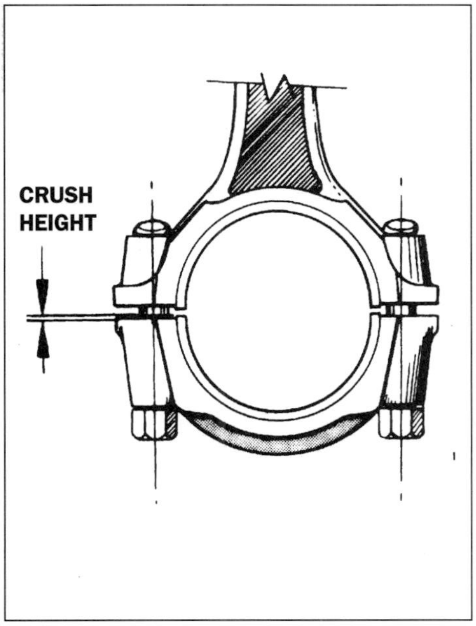

The amount of projection of the bearing above the bearing-bore of the piston is called the crush height. When the cap is tightened these ends squash together preventing the bearing from spinning in the bore.

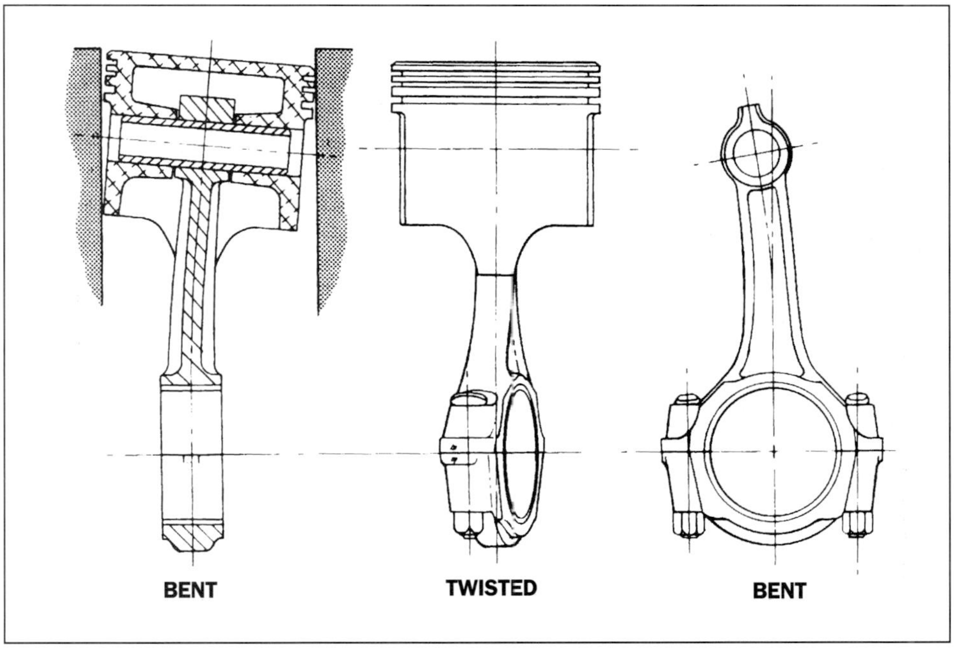

This drawing shows the three ways a rod can be bent. Besides the three ways shown here they can also be bent or twisted in both directions at once. Generally, it's best to replace a bent rod even though it is possible to straighten it.

pressure and the "idiot light" failed to come on or came on after the engine seized. So we know it happens. Therefore, you must pay careful attention to this part.

You can learn a lot about your oil pump by looking at the bearings. How were the rod and main bearings? Were there visible signs of dirt or metal particles having scored the bearings? If there were, you'll know without looking. Those bits of debris did their dirty work to the pump before they did it to the bearings.

The first test is to see if the oil pump will pump. Immerse the pickup in a pan of solvent and turn the rotor by hand. After two or three revolutions, solvent should gush out of the pump's pressure port. This shows the internal clearances are near or within tolerance, and the rotor will turn freely.

The next step is to disassemble the pump for visual inspection. Remove the two remaining bolts holding the cover to the body. Be careful lest the rotors fall on the floor. You can expect little wear from normal usage as this unit is so well lubricated. This won't be the case, however, if dirt or debris got in. Remember, the oil pump picks up unfiltered oil. The following specs are for all "B" and "RB" engines.

Check the clearance between the housing and outer rotor with a feeler gauge. It should be less than 0.012-in.

Check rotor end play by placing the cover plate over half the housing, then measure the clearance between the plate and the rotors with feeler gauges. Maximum clearance is 0.004-in.

Rotate the pump until the tips of the inner and outer rotors are at their closest point. Clearance between them should be less than 0.010-in.

Turn the pump upside-down and let the rotors fall into your hand. Be very careful not to drop or scratch them. Check for scoring, pitting, deep grooves or any signs of wear from foreign matter.

Inspect the housing for any damage, particularly scoring or cracks. Don't put the rotors into a worn or damaged housing. Pay particular attention to the inlet and outlet ports in the housing. You may find foreign particles in here. If you have your micrometers handy, check the inner and outer rotor thickness and diameter.

On all "B" and "RB" engines the inner rotor should be no less than 0.942-in. thick; the outer rotor no less than 0.943-in. Outer rotor diameter is 2.469-in.

Rebuild kits can be purchased from TRW/Federal Mogul, as Part No. 51047, and will work for all years and all models. Complete aftermarket oil pumps may also be purchased as Part No. 50033 for all engines except the 426 Hemi. Order this as Part No. 50047. Unfortunately, Chrysler does not manufacture an oil pump rebuild kit. Mopar Performance, Chrysler's high performance parts and marketing group, only offers a racing type pump.

The new piston bore should also be polished and deburred.

When the wrist-pin bore reaches a temperature just before it begins to glow, it's pulled from the oven, centered in the piston and the wrist pin pushed through. The fixture in which the piston sits has been set to limit the travel of the wrist pin, centering it correctly within the piston.

The last step is to be sure the connecting rod is square with the piston.

Pressure Relief Valve

The only check to make with the pressure relief valve is to determine if it is free. Bend a piece of coat hanger wire into a small hook. Use this to reach in and attempt to move the plunger. If you can, all is in order. If you can't, you must free it. Mount the oil pump body (with the square bolt-head in the up position) in a vice with a piece of wood to protect the face. Remove the spring retainer cap (the square bolt-head) with a 5/8-in. end wrench. Inside will be the spring and plunger. Remove the spring and again try to push the plunger from its backside with a piece of coat hanger wire.

If you have no success with this, soak the assembly with lacquer thinner or carburetor cleaner until it frees up. If it never does, you'll have to replace the pump.

If you get it out it can be cleaned with thinner or carb cleaner while scores and pits can be removed with 400-grit wet-or-dry sandpaper.

Before reassembling the oil pump test the spring. Spring free length should be 2-9/32 to 2-19/64-in. and should test 14.85 to 15.85-lbs. when compressed to 1-19/32-in. The 426 is the same except it should test 22.3 to 23.3-lbs. when compressed.

When you reassemble the pressure relief valve, don't forget to lubricate the piston and bore.

Reassemble the Oil Pump

Before reassembling the oil pump, be sure the oil seal ring is in good condition and the parts have been thoroughly cleaned and lubricated. Install the cover and torque the bolts to 10 ft-lbs. Check your work and prime the pump by immersing it in clean oil and turning the rotor until oil comes out.

SUMMARY

I hope I've stressed cleanliness until you're tired of hearing about it. If so, it should start to be second nature to you and you won't have to think—you'll just do. One metal filing can mean starting all over from the beginning. So keep things clean and only do the job once.

Now, your block, pistons, cam and crank are all reconditioned and ready for assembly. Be sure everything is oiled with WD-40 or one of the other water-dispersing oil sprays and wrapped in plastic bags. Stand the crank on its flange end and wire it to the garage wall so it won't fall. Do the same with the cam or hang it from the rafters, just don't let these two parts lie on the floor. If done for an extended period of time they'll warp, exceeding maximum run-out.

In the next chapter we'll get the heads ready and then we'll put this engine back together. ■

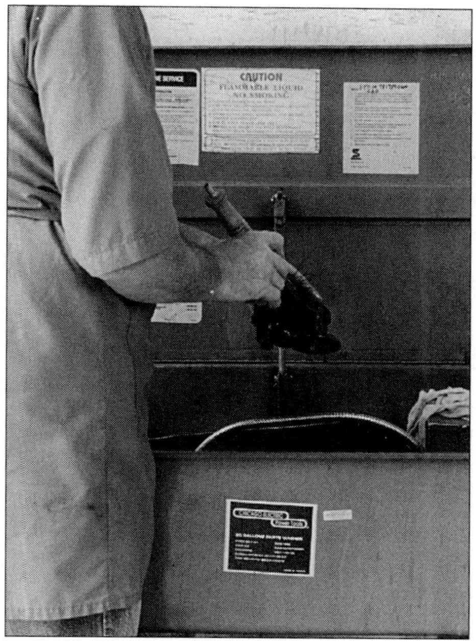

Clean the oil pump well before you open it to take measurements. It helps keep out foreign matter.

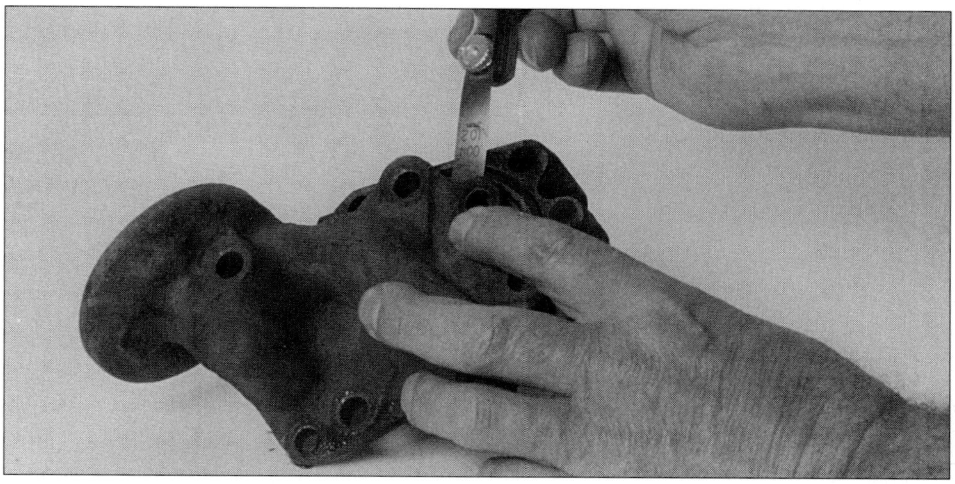

Maximum clearance between the cover plate and the rotor is 0.004-in. Again, be sure there is no foreign matter between the two pieces.

Maximum width between inner and outer rotors should be no more than 0.010-in.

Although this rotor is a bit scored from all the debris it has ingested over the years it still looks like it has plenty of life left in it.

RECONDITIONING THE HEADS

6

Once again you're faced with a decision: should you do all of the disassembly and assembly of the heads or turn the job over to the machine shop? The cost is about the same. Unfortunately, you'll save little money doing the dirty work. The real problem is the machine shop's guarantee. In most shops if you do the assembly work, they won't guarantee the finished product. I feel it's important for you to do the disassembly so you can check for wear and breakage as you go along. If you've done head work before and feel competent about the machine shop, then do the final assembly yourself. If you feel the need of a shop guarantee, let them do the work.

You're not off the decision hook yet. If you find the head is damaged or worn beyond repair, you'll have to exchange it across the counter for a rebuilt one or find an acceptable one from the salvage yard. Most salvage yards guarantee their parts—not to work but to be replaced if they're not satisfactory. Therefore, if you choose to buy a used head you may find yourself going back and forth to the salvage yard until you find just the right one.

Please understand I'm not trying to discourage you or push machine shops. I simply want you to be armed with the facts before you begin your project. Whatever you decide to do you'll find everything you need to know, and a bit more, in the following pages.

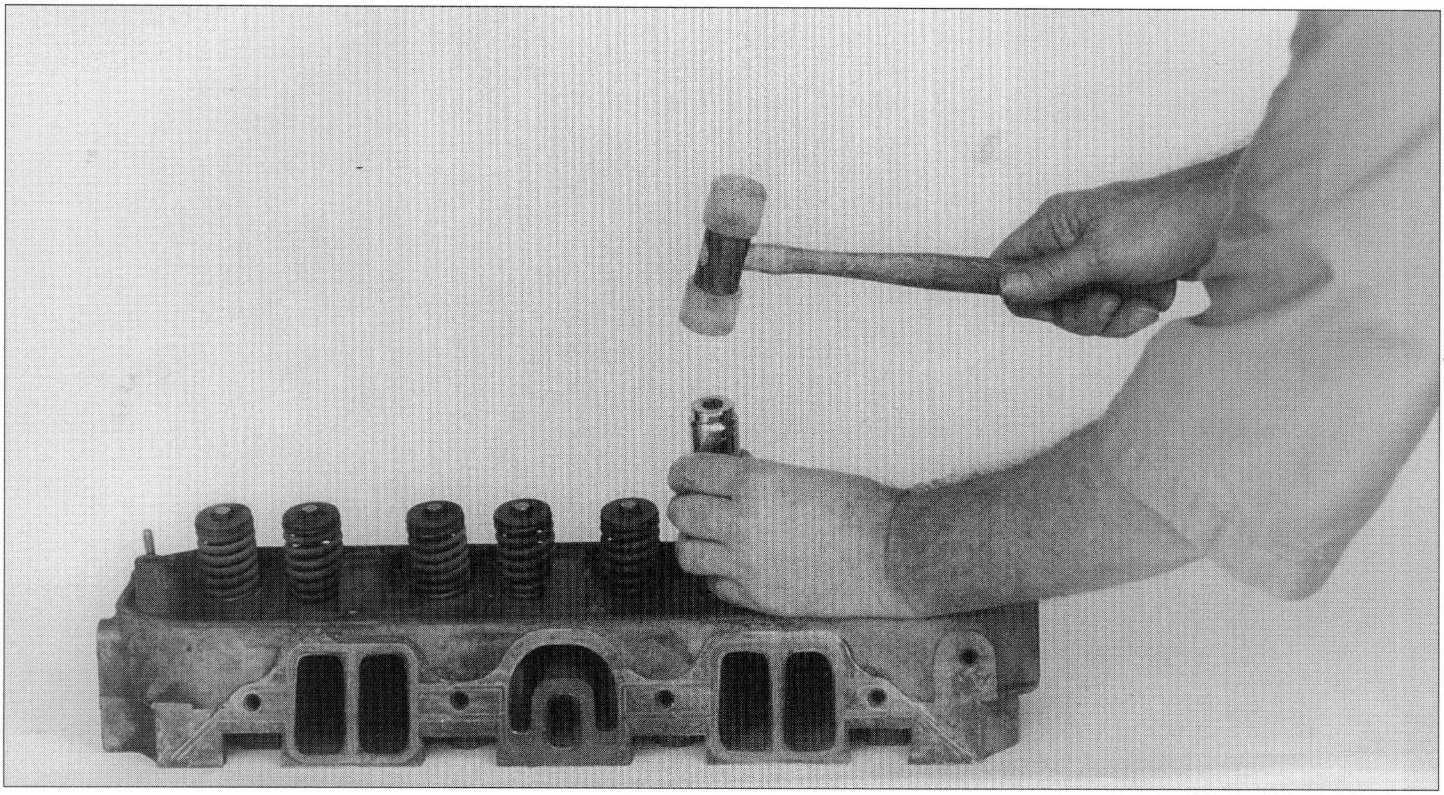

This is the "hit and run" process described in the text. It's a fast way to remove the keepers. Sometimes, however, the keepers go flying away from the percussion and you have to run all over the shop looking for them.

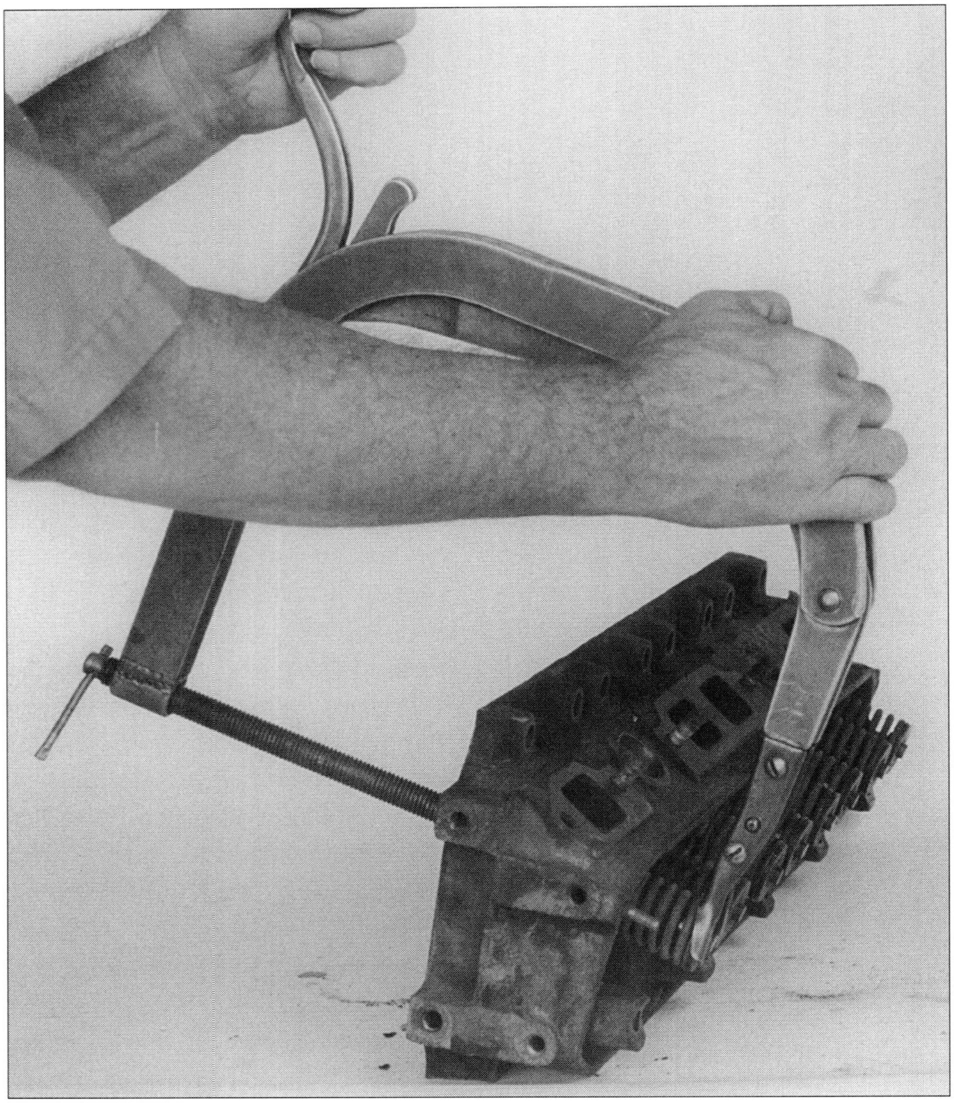

Use a spring compressor to get to the keepers if you want to be sure of "keeping" them.

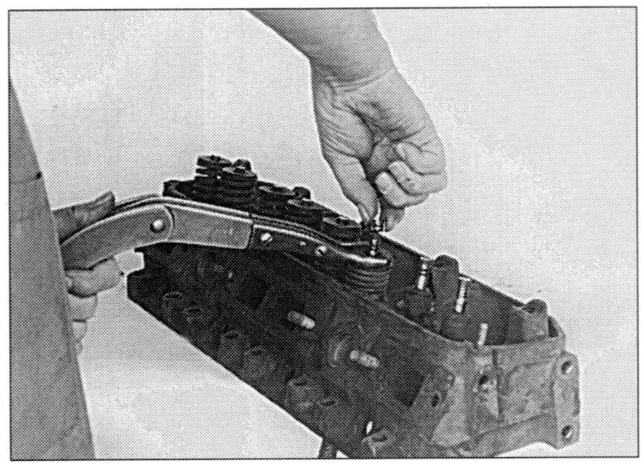

The valve keepers come in two parts. When the spring is compressed they may be removed by hand. Note there are two-groove valve stems and four-groove valve stems. The four-groove valves are exhaust and the others are intake.

DISASSEMBLY

I hope you removed the rocker arm assembly before you removed the head from the block as demonstrated in Chapter 3. If you failed to do so, remove the rocker arm assembly now, keeping the arms in their original order. Each assembly should be marked right or left (the driver's side is always left). Set them aside and we'll work on them a bit later.

To make the job a little nicer, wash the heads with solvent or degreaser before you begin the disassembly process. Later, you'll get the heads as clean as you did the block in the previous chapter. Now, find a small box for the parts and begin removing the valves.

Valves

The most common way to remove the valves is with a valve spring compressor. It looks like a giant C-clamp. The valve spring compressor holds the valve closed and compresses the spring with its retainer to expose the valve keepers. Remove the keepers, release the clamp and remove the retainer, spring, valve-stem seal and valve.

An alternative method to the valve spring compressor release is the "hit and run" method. Set the head flat on the bench. Place the nut end of a 5/8- or 9/16-in. deep-well socket over the valve-spring retainer. Give the base of the socket a sharp rap ("hit") with a good size hammer. The blow should compress the spring and allow the keepers to fall away from the valve-keeper grooves. Sometimes they get away from you. Then you must "run" around trying to find them. Usually, however, they fall onto the top of the head and stay there.

Once all the valves are removed, remove the valve-stem seals. They will pull off the spring-end of the valve guide. Don't be surprised if they're broken or missing. It's their nature. Remove and discard all of them as you'll be replacing them with new ones.

Often the valve will hang up at the area of the keeper grooves as you slide it out. This is caused by burrs at the edges of the grooves. Use a very fine file to go around the keeper grooves to remove them. The valve should then slide out smoothly.

These are the four parts that make up the spring assembly; spring, spring retainer and two valve-spring keepers. Try not to lose the retainer or keepers. They can be used again and are expensive to replace.

This is what the valve seals look like after years of service. Often they will be completely missing. Never try to reuse this part even if you can.

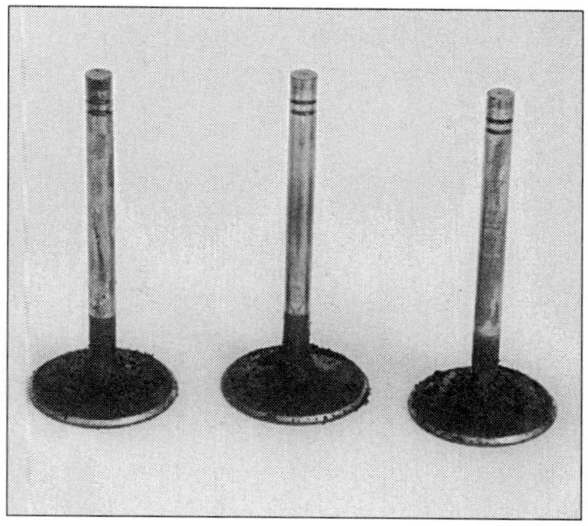

This carbon buildup on the intake valves is probably a result of the worn valve seals. Other than that, these valves look to be in good shape.

If you want to reuse your valves, and you will if you check the price of new ones, keep them in order as they're removed. Unless you install new guides, the valves must go back into their original positions. Standard practice for as long as I've known it is to drill 16 holes in a board, number the holes 1-8 passenger and 1-8 driver, and insert the valves as marked. This worked well in my case for many years. Then one day someone knocked my board off onto the floor, full of valves. They all spilled out leaving me no alternative but to buy new ones. Now I number them with little masking tape flags.

Core Plugs

There are two core plugs (freeze plugs) on each head. These should be replaced even if they look good. Remove them in the same manner as those in the engine block. Use a punch to rotate the plug, then pull it out with a pair of channel-lock pliers using the heel of the pliers as a fulcrum.

Cleaning & Inspecting

There is no such thing as clean enough. It's either completely clean or it's not. In the engine rebuilding business it must be clean. Begin with the gasket surfaces and work them clean to the metal. Get out a screwdriver and scrape out any carbon buildup inside the combustion chamber. Be careful not to gouge the floor of the chamber, however.

After scraping, finish off with a wire brush—preferably one you've chucked into a drill motor. Work around inside the combustion chamber getting out every trace of carbon. Then, sand the head-gasket surface with 80-or 100-grit sandpaper wrapped around a flat wooden block.

Next you'll check the head for cracks. Look carefully around the valve seats in particular. Are there any cracks between an intake and exhaust? If you find any cracks the head should be discarded. The very best way to find cracks is just as you did with the block: have it Magnafluxed. This costs little and often turns up cracks invisible to the eye. Although small cracks can be welded, usually the cost of welding and machining is about the same as a good used head.

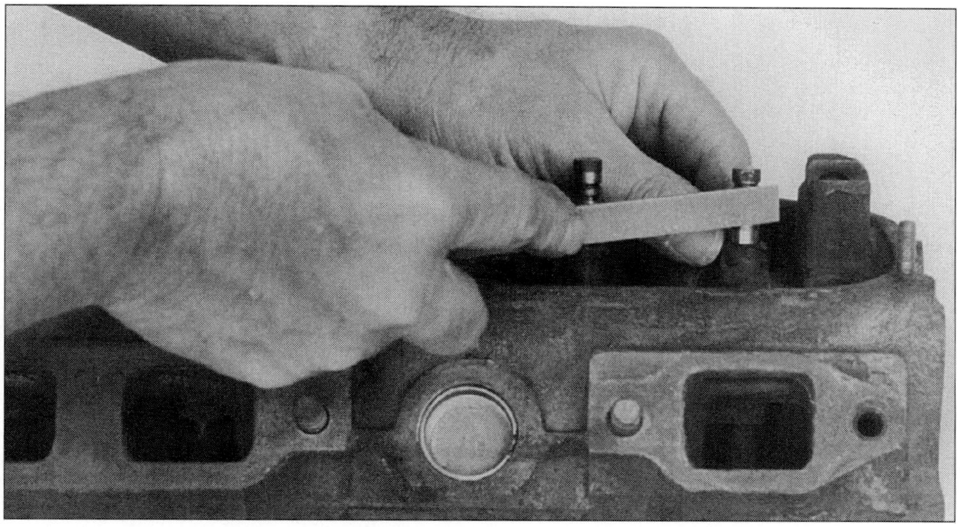

The constant hammering of the keepers against the valve stem groove edges "mushrooms" them. This prevents the valve from sliding out of the guide. A few strokes of a very fine file will remove this burr and allow the valve stem to pass through.

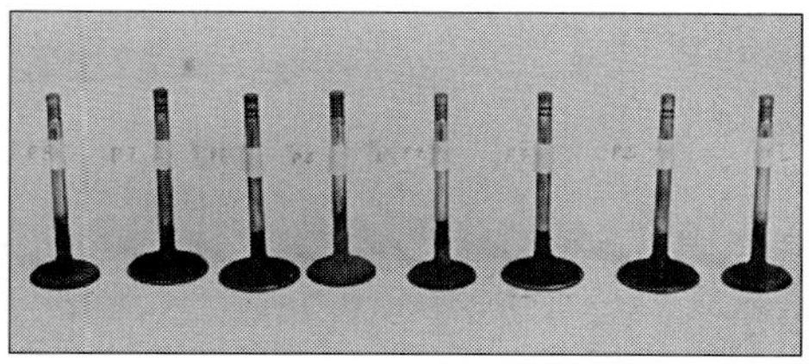

Don't laugh at my flags. Short of falling into the solvent tank I'll always know which one is which!

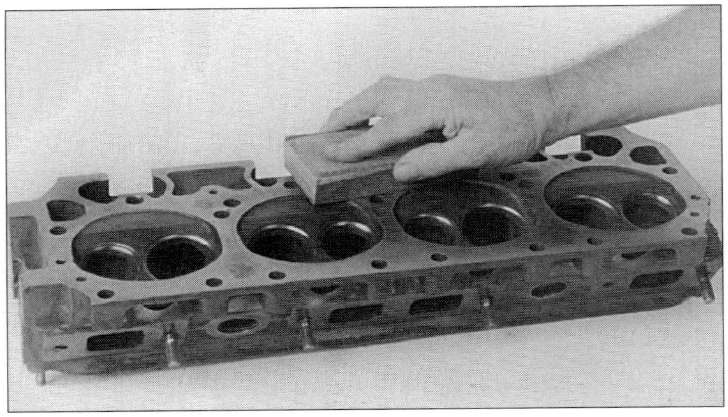

Begin your cleaning, as always, by scraping and sanding the gasket surfaces. Later, this will allow accurate measurements of the head for possible warpage.

Finish your cleanup by washing the head in solvent or degreaser. Then, get out the old rat-tail file and go to work on the waterjackets just as you did on the block. These must be cleaned to allow for the free flow of coolant. If you miss this step you run a serious risk of cracking or warping the heads from heat buildup.

FIXING STRIPPED STUDS OR BROKEN BOLTS

Because of very high temperatures associated with the exhaust manifold, severe oxidation—rust—occurs on the manifold and its associated bolts. If your engine has been sitting around for many years without the manifold having been removed, it's very likely that this oxidation process will have attacked at least one or two of these bolts. Usually, this results in breaking off the bolt above the threads when you remove it. In the case of our project car there was no bolt at all in the forward hole as there were no threads to hold it!

If you suffer broken bolt or stripped thread syndrome, not to worry. The doctor is just down the street at the parts house with medication in the form of a product called a Heli-Coil. A Heli-Coil is simply a new set of threads to replace damaged ones.

To use this nifty device, determine the thread size, pitch and depth of the hole that needs to be repaired. For our demonstration, I needed a 3/8-in. N.C. (National Course) X 1/2-in. deep set of threads. If you're doing this job, you'll need the same size Heli-Coil kit and the corresponding 25/64-in. drill bit. The drill bit size is important and is stamped on the Heli-Coil kit. If the bit is too small you run the risk of damaging the tap or the head. If it's too big the resulting hole will prevent a deep enough thread for the coil to bite into.

The kit contains three items: a tap, an insertion tool and several Heli-Coil inserts. Some kits come with a fourth item, the correct drill bit. These three or four pieces ar a matched set and must be used as such.

Begin the repair by drilling out the damaged threads. If you're working near a waterjacket take extreme care not to penetrate it with the drill. Remove only the threaded portion of the hole. A piece of masking tape around the drill bit will act as a depth gauge.

Next, cut a new set of threads with the tap provided. The new threaded hole will be much larger than the original. This is to allow for the insertion of the Heli-Coil.

Thread the coil onto the insertion tool, making sure the tang on the bottom of the coil fits into the groove at the end of the insertion tool. Thread the coil into the new threads. When the coil bottoms in the new threads, the tang will break off to lock the coil in place.

If you cut the threads too deep, insert the coil until it's just below the surface of the head, then remove the insertion tool. Select a punch just a bit smaller than the ID of the coil. Insert it until it touches the tang of the coil, then give it a whack with a hammer. This too, will break off the tang and leave the coil well placed.

Heli-Coil repairs work well on any thread problem, even damaged spark plug threads. If you can't repair the threads with a tap or chaser, reach for the Heli-Coils.

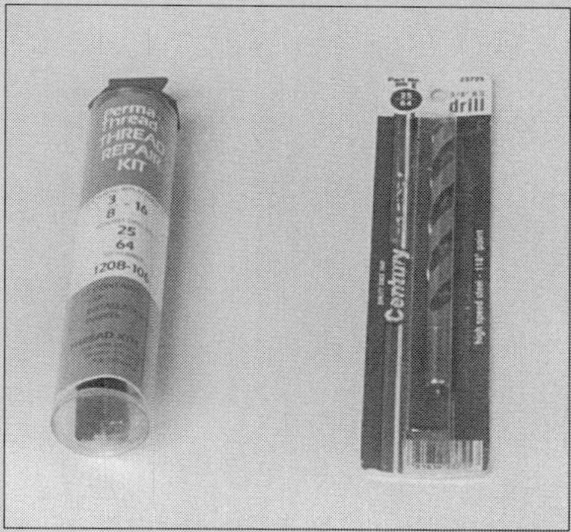

A complete Heli-Coil kit includes the correct drill bit and tap.

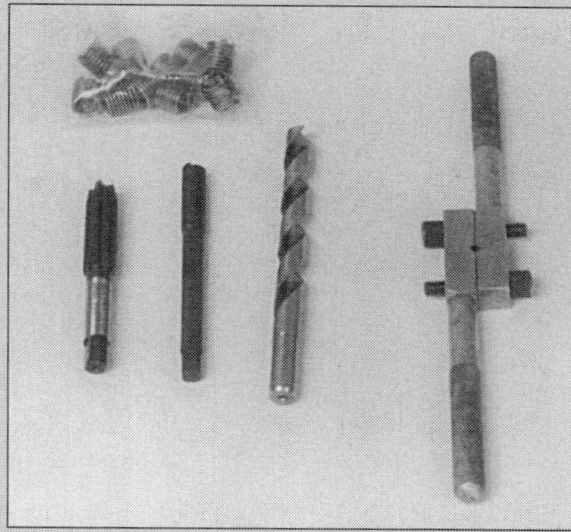

This and your drill motor are the only tools you'll need for the job. Note that the insertion tool will fit in the tap handle.

Here's our poor head with completely stripped threads. In less than 5 minutes it will be as good as new.

Begin by drilling out the damaged thread.

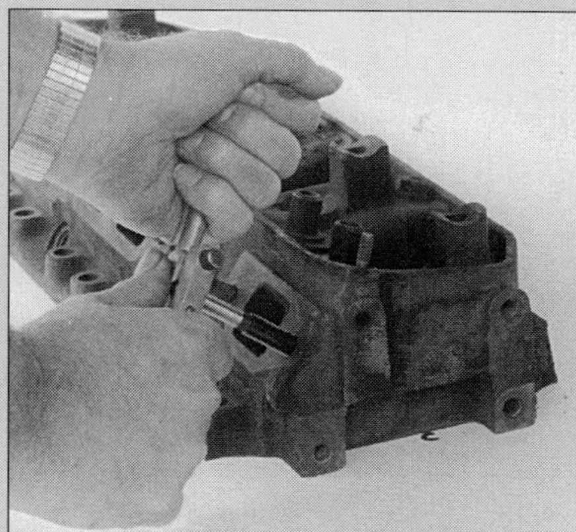

Tap the new hole to the outside thread size of the Heli-Coil.

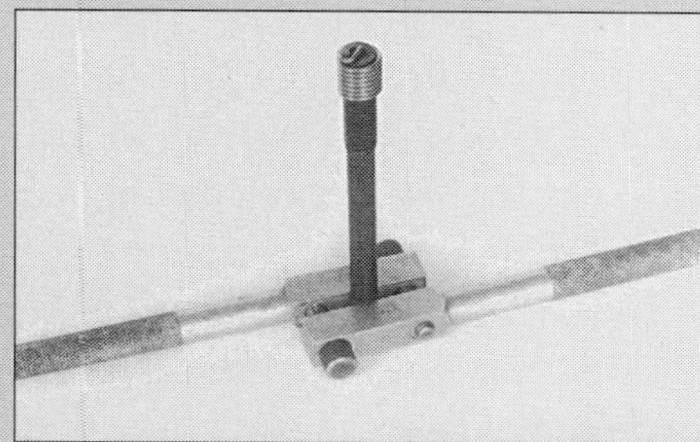

Here's the close-up of the coil threaded onto the insertion. Note the tang on the coil and the groove in the tool.

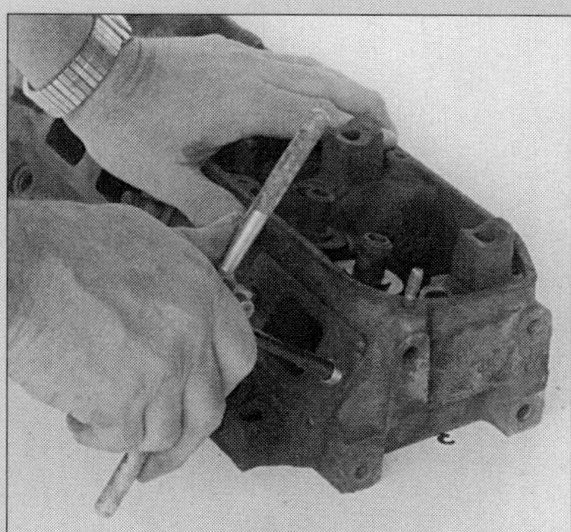

Thread the coil into the newly tapped hole. The tang of the coil will break off to lock the coil in place.

The newly repaired thread.

Use whatever it takes to open the water passages. Here, I'm using the tang of a file.

Head Surface Flatness

Everything we did to the block we must do to the heads. This includes checking for flatness as you did on the block deck. The sanding you did on the head gasket surface removed any tiny carbon or metal particles that could give a false reading when you measure for flatness. Although smaller than the block, the cylinder heads receive the same combustion load. Consequently, they're highly subject to warpage, resulting in coolant or combustion leaks. Either of these can ruin an otherwise good head. To avoid these problems, the heads must be checked carefully for flatness.

Most early Mopar big-block engines were built using steel-shim head gaskets. Remember, this was in the day of high octane (100+) gasoline. The steel shim gaskets were strong and helped keep the compression ratio up. Today, with little to no lead in our gasoline it's impossible to find anything better than 92 octane at the pump. Therefore, you'll probably want to use a composite head gasket when you assemble the engine. This will lower the compression, allowing you to burn the lower octane gases available today.

All of the above is said to help you determine the maximum warpage allowable for your heads. If you choose to use composite gaskets, the allowable warpage is .006-in. Using steel shim gaskets this figure is reduced by half to .003-in. The lower figure for the steel shim gasket is a result of that gasket's limited gap-filling ability as compared to the composite gasket. The steel shim gasket has virtually no "crush factor." Therefore, the head must be perfectly flat in order to seal properly.

Mopar Performance (formerly Direct Connection) has foreseen the need for a good composition gasket. Order part #P4349559. This gasket fits all "B/RB" engines. It features a special O-ring gasket for improved compression sealing. Compression thickness is .042-in.

Get out your straight-edge or framing square and feeler gauges. Lay the straight-edge across the head gasket surface lengthwise first, then crosswise. Attempt to insert either the .006-in. feeler gauge or the .003-in. feeler depending on which gasket you'll be using. If you can squeeze anything bigger than a .006-in. gauge in anywhere between the head and straight-edge and you plan to use composite gaskets, you'll want to have the head milled (resurfaced). Likewise, if you can insert anything larger than .003-in. between the gasket surface and straight-edge and you plan to use original steel shim gaskets, that head will have to be resurfaced.

Cylinder Head Milling

The head can be resurfaced (milled) the same way as a block deck. Please take note: if one head must be milled, the other should be milled the same amount. Milling only one head will create a difference in compression ratio between one bank of cylinders and the other. This is also true of the block deck, but the

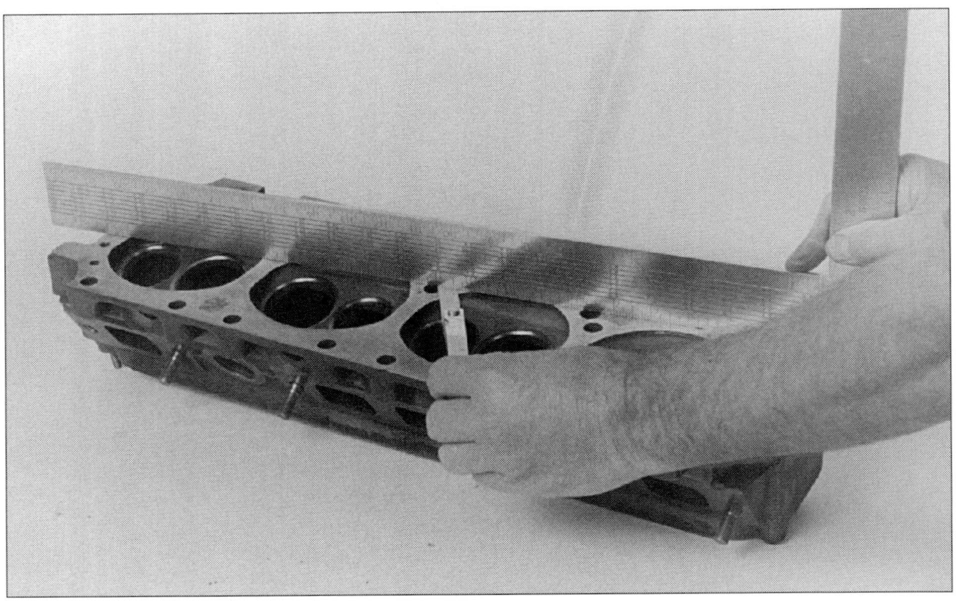

Checking warp with a straight-edge laid across the full length of the head. Gaps are measured with a feeler gauge. The distance between the head and the straight-edge should be less than .006-in. Our project head passed with flying colors.

valve seat and face.

If excess wear is found, new valve guides or valves can be installed, or even a combination of both. So, with this in mind, let's take a look at how to measure s/b valve guide-to-stem wear.

For all B/RB engines except the 426 Hemi, Chrysler recommendations call for 0.001-0.003-in. stem-to-guide clearance for the intake valves and 0.002-0.004-in. for the exhaust valves. For the 426 it is 0.002-0.004-in. for the intake and 0.003-0.005 for the exhaust valves. There are several ways to measure this clearance, varying in accuracy and convenience.

Measuring Valve Guide Wear

Let's begin our determination of valve guide wear with the simplest and easiest method. It's called the "wiggle test." If the valve stem "wiggles" within the guide, it's worn—simple as that.

The accepted way to perform the test is to place the valve in its guide until the tip of the stem is flush with the top of the guide. This places the valve on an unworn portion of the stem, so the measurement accurately reflects guide

INTAKE MANIFOLD VS. CYLINDER HEAD MILLING

REMOVED FROM HEADS	REMOVED FROM MANIFOLD SIDES	ROUNDED OFF
.001	.00123	.0012
.002	.00246	.0025
.003	.00369	.0037
.004	.00492	.0049
.005	.00615	.0062
.006	.00738	.0074
.007	.00861	.0086
.008	.00984	.0098
.009	.0110	.0110
.010	.0123	.0123

If you have your heads milled, use this chart to determine how much to take off the intake manifold.

machine shop will automatically mill both deck surfaces.

Milling the heads creates yet another problem. This is the relative position of the intake manifold. Whenever a head is milled, the intake ports move downward and in, relative to the intake manifold. So, for every 0.010-in. removed from the head surface, 0.0123-in. should be removed from the manifold sides. For example, a cut of 0.020-in. from the head surface would require a corresponding manifold cut of 0.0246-in.

0.020-in. is a large cut and used only as an example. Limit any milling operations to under 0.010-in. Again, it's the problem with low octane fuels. Anything over 0.010-in. will raise the compression ratio to such an extent that detonation (preignition) will occur.

VALVE GUIDES & STEMS

You may think of valve guide-to-stem wear just as you do piston-to-bore wear. Both types of wear cause an increase in oil consumption and a loss in performance. Piston-to-bore wear lets oil and compression pass between the piston and bore. Valve guide-to-stem wear lets oil and exhaust gases pass between valve and guide.

A badly worn valve guide won't guide the valve squarely onto its seat. Consequently, the valve will wiggle and slide onto its seat and won't seal properly. This will eventually destroy the

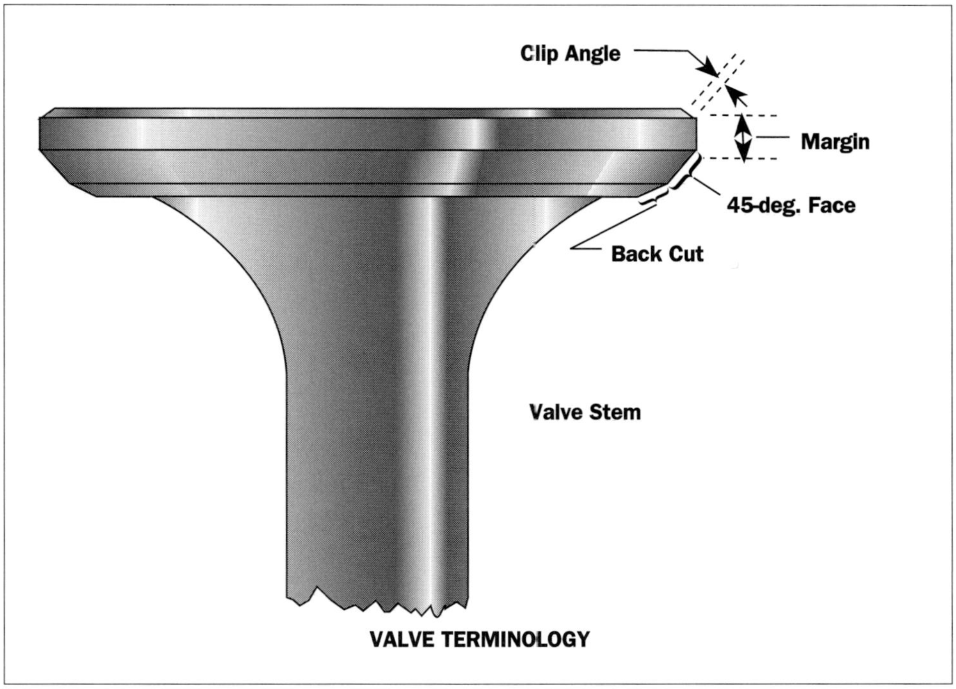

Valve nomenclature. The two grooves at the bottom of the stem indicate this to be an intake valve.

wear, not guide-to-stem clearance. The valve is then moved, i.e. “wiggled” from side to side.

The amount of movement from one extreme to the other can be measured with a dial indicator mounted 90° to the valve head, or the depth-gauge end of a vernier caliper. This measurement is then divided by 3.5 and the result is the approximate valve-guide clearance. As an example, if the outer limits of wiggle are 0.009-in., dividing this by 3.5 gives you 0.00257-in. within the limits for either intake or exhaust.

This trick gives you approximate clearance. If your figures are on the borderline, double-check with one of the following methods or play it safe and consider the clearance excessive. With all the measuring methods, including the wiggle test, be sure to measure clearance in two directions 90° to one another, or front-to-rear and side-to-side.

The most accurate way to measure stem-to-guide clearance is with a small-hole gauge and micrometer. A small hole gauge works something like an inside mike. It expands to fit the hole you’re measuring. With the small hole gauge, measure around the guide at its top and bottom to find its maximum diameters. Remove the gauge and measure the ball end with an outside micrometer.

My long-time friend Tim Mints selects a cutting tool and reams out the old guide.

The valve stem is then checked and this measurement subtracted from the measurement to find the clearance. If the clearance limits are exceeded, the guides should be reconditioned.

Chrysler recommends checking guide wear with a dial indicator. Install a valve in the guide about 1/8-in. off the seat. Mount the dial indicator 90° to the valve stem in the direction you want to measure wear, contacting the valve stem as close to the top of the guide as possible. Pull the valve away from the tip of the indicator and zero the dial. Now, push the valve toward the indicator and read the clearance directly.

Amount of Guide Clearance—In spite of Chrysler’s recommendation of 0.0045-in. as maximum stem-to-guide clearance, we’re going to use the standard figure most machine shops consider the maximum allowance, 0.004-in. When you consider new intake-guide stem tolerance to be 0.001 to 0.003-in., then 0.0045-in. is getting pretty large. I believe if your engine had this much wear you’d have been burning a lot of oil!

Maximum clearance is a bit misleading. If you found your intake guide clearance to be 0.004-in. or almost 0.002 to 0.003-in. over new you’d say, “Well, that’s fine. I’m within limits.” Yes

SOME THOUGHTS ON METAL WEAR

Suppose you found intake guide clearance to be 0.004-in. or about 0.002 to 0.003-in. over new. “Well,” you say, “this is within maximum limits.” Well, yes it is, but it leaves very little room for additional wear. You could reassemble and install the engine, drive the car around the block and “wear out” the guides.

Additionally, there is a phenomenon concerning wear—the more wear on the part, the faster the part wears. It works something like this: if the guides wear 0.003-in. in 75,000 miles you might think they wore 0.001-in. every 25,000 miles. This is not the way it happens.

Using approximate figures, the guides probably wore 0.001-in. in the first 40,000 miles, the second 0.001-in. in the next 20,000 miles and the last 0.001-in. in the last 15,000 miles. Where then, does this leave guide clearance in terms of remaining mileage with only 0.001 or 0.002-in. left to go? As I said above, it may only get you around the block.

Keep this in mind as you measure all worn parts. When you approach the limit, think in terms of how soon you’ll exceed the limit.

Using yet another cutter, Tim makes a "bellmouth" cut to allow the new guide to slide in without binding.

you are but you've no room left for wear beyond that point. You could install the old valves and in a 1,000 miles be worn beyond the limits (see sidebar on page 98). For your own benefit, as long as the engine is disassembled, if wear is over the tolerance listed for new, then go through the complete reconditioning process.

Valve Guide Reconditioning

There are three ways to correct excessive valve-guide clearance: better-than-new, satisfactory and "don't tell anyone I did this!" We'll look at them in that order.

Valve Guide Inserts—For better-than-new, consider valve guide inserts. Bronze inserts are the best. You can easily expect 150,000 miles before you have to redo them. The next level is the same material as the head, cast iron. These will give you the same performance as the originals. The difference in price, though considerable, is not out of reason. Where possible choose bronze over cast iron.

Although there are tools on the market that allow the home craftsman to do his or her own inserts, these should be avoided. There are too many chances for you to damage the head. Leave the reaming of the guide, installation of the guide and rereaming to the machinist. You can assemble the head yourself when the machine work is finished. The savings are not worth ruining the head.

Oversized Valves—Using valves with oversized stems is fine if most of the valves must be replaced. If they're burned beyond use or bent, then they must be replaced. If this is the condition of your valves then just have the guides reamed to fit the stems of the new valves. New valves are available in 0.005, 0.015 and 0.030-in. oversize. Under these conditions, needing all new valves, reaming to oversized is fine. But, if the valves are okay, you'll be money ahead reconditioning the guides.

Knurling the Guides—Just as you can knurl pistons to reduce piston-to-bore clearance, you can knurl valve guides to take up stem-to-guide clearance. A knurling tool is run through the guide to reduce the effective size of the guide hole. The guide is then reamed to fit the new valve stem.

With an impact driver the new guide is forced into the head.

The new guide is now reamed to the exact size needed for the new valve stem. A reamer can finish a guide to an accuracy of four decimal places, or within 0.0001-in.

When new guides are installed, the valve seats must be ground to perfectly recenter the valve in the head. This must be done whether or not the valve seat is in good condition.

Knurling produces a series of peaks and valleys. It is the pushing of the material from the "valleys" that produces the "peaks" and reduces the size of the bore. A knurled guide then gives about 50% less bearing surface than a new guide and will wear about twice as fast—or worse.

With knurling you get what you pay for: very little. You won't solve the problem with this approach, you'll only postpone it.

Valve Seats

Valve seat reconditioning is a necessary part of every rebuild, particularly if new guides are installed. When new guides are installed, it is impossible to center them perfectly with the valve seats, consequently, the new seats must be ground concentric with the valve guides. A mandrel or pilot centers in the valve guide to ensure the seat is ground concentric with the guide.

Grinding the Seats—A new valve seat consists of three cuts made with an electric or pneumatic grinder. The top and bottom cuts are made with 30° and 60° stones. The final cut is made using a grinding stone with the face at a 45° angle to the valve guide. The final cut at 45° is the most important. This is where the valve face makes contact with the seat.

The other two cuts, the 30° top cut and 60° bottom cut establish seat width and inside and outside diameters. The top cut trues up and determines the outside diameter of the valve seat. Valve seat OD should be ground approximately 1/16-in. smaller than the valve OD. Valve seat width is established by the 60° bottom cut.

On all of their big-block engines, Chrysler recommends cutting the intake valve seat to a width of 0.062 to 0.090-in. (1/16 - 3/32-in.) and the exhaust valve seat to 0.046 to 0.062-in. (3/64 - 1/16-in.) Some machine shops and performance enthusiasts will recommend a larger exhaust valve seat width to provide better heat transfer and reduce "mushrooming" of the seat. Maximums, however, should not exceed 0.090-in. intake and 0.100-in. on exhaust.

The 426 Hemi requires a further step. When the valve seats are reconditioned on the Hemi, the seat width reduction requires special stones be ground to

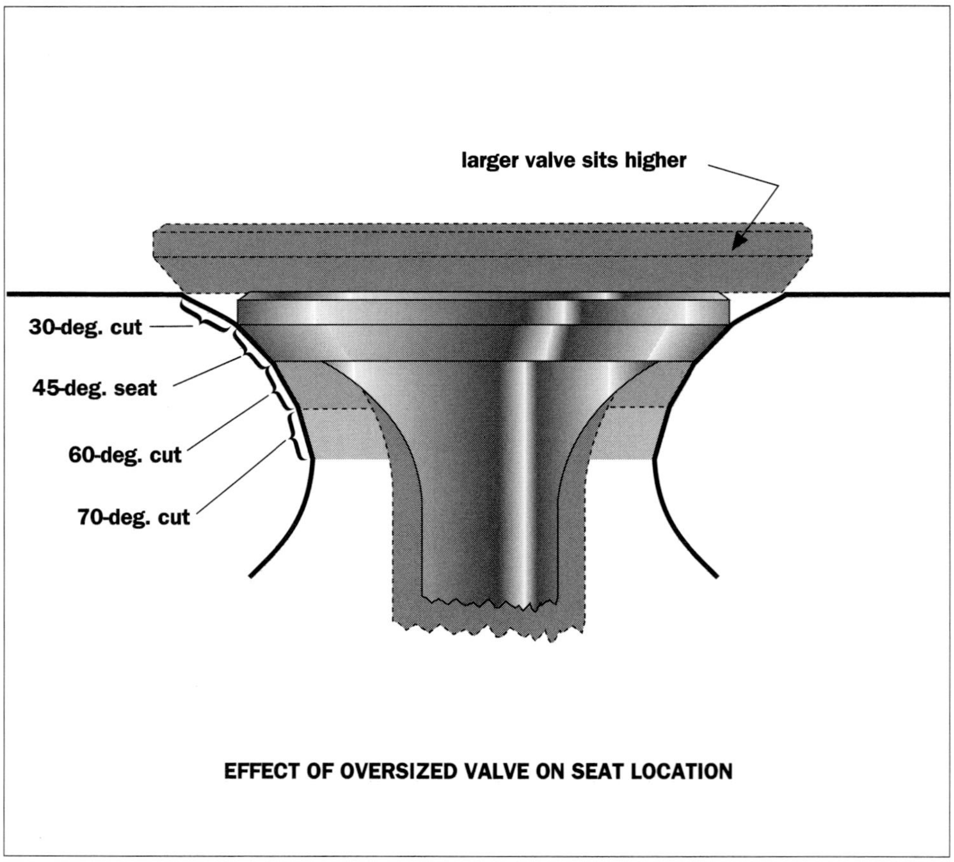

This illustration shows the three cuts made in the head for a proper valve seat. The first cut is made at the top and is 30° to the centerline of the valve. The second cut is at the bottom and is 60° to the centerline. Finally, the 45° face cut is made.

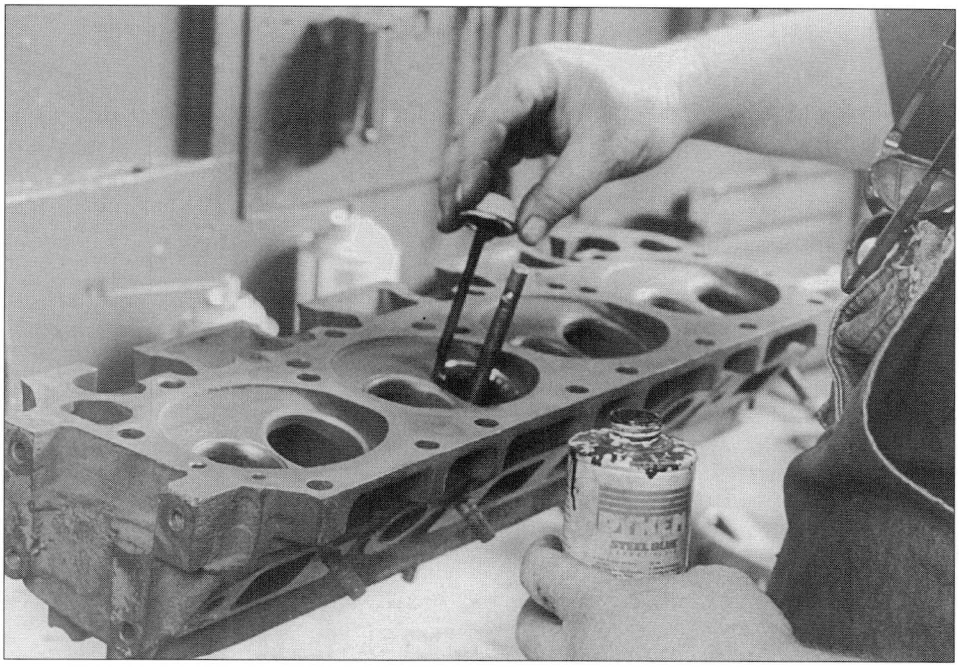

Tim applies a blue dye to the surface of the area to be cut (ground). This makes the new cut more visible so it can be checked for voids.

maintain the radius from the seat approaches to the seat. The intake requires a .244-in. radius and the exhaust must have a .261-in. radius. The drawing below from Chrysler Service Bulletin No.64-73 further clarifies this.

Finally, the concentricity of the seat should be measured. This is measuring to find how close to perfectly round the seat is. Suppose the valve face was perfectly round while the seat was a bit egg-shaped. Obviously, valve and seat would not seal. Concentricity, or roundness, is measured with the dial gauge and should not exceed 0.003-in. You've worked so hard on these heads, however, that you should really shoot for a seat that is no less than 0.002-in. egg-shaped.

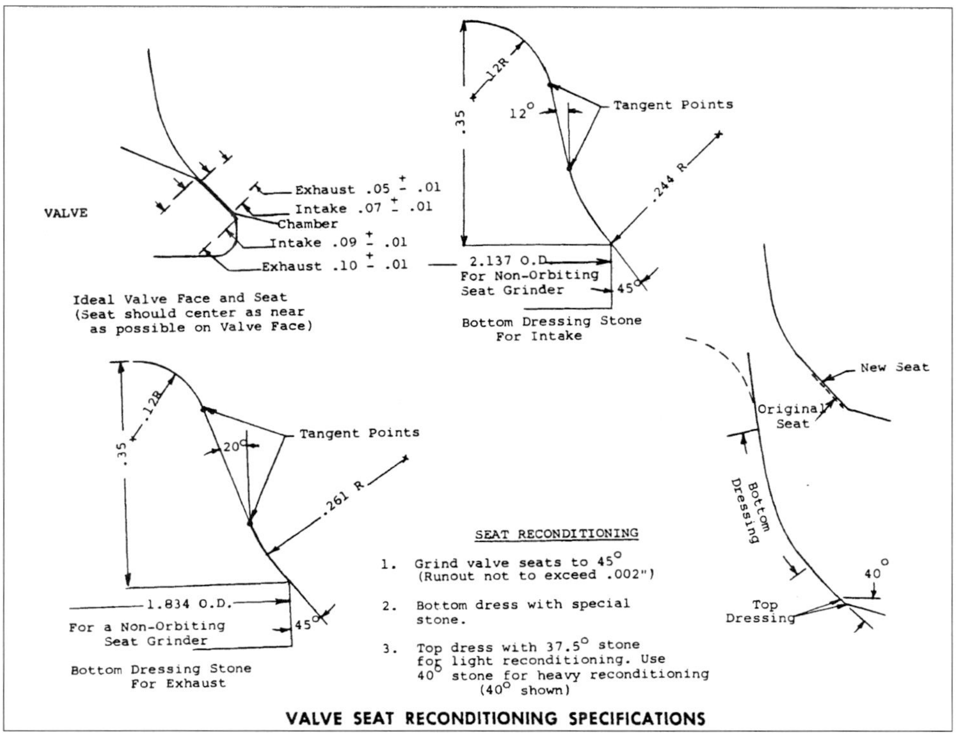

These are factory valve seat reconditioning specifications for the 426 Hemi. Your machine shop should use this chart to dress the stones used to cut the seats.

VALVE INSPECTION & RECONDITIONING

Did you find burned valves when you disassembled the heads? Were any bent? These problems are usually easy to spot. Harder to spot though is valve stem wear. The stems wear just as the guides, only slower. To determine the amount of wear you'll have to get out the micrometers again and take some measurements.

Measuring Valve Stem Wear

The nominal, or standard size, for all big-block intake valve stems is 0.372 to 0.373-in. and 0.371 to 0.372 for the exhaust valve stem. To determine the amount of wear, measure the valve stem in an unworn area. Then, measure again in a worn area and subtract the smaller figure from the larger to get the amount of wear.

The greatest wear usually occurs at the tip end of a valve's stem-to-guide contact surface. This area is easily recognized as the shiny portion of the valve stem. There's a sharp division between it and the unworn surface.

Maximum valve opening is represented by the end of the shiny surface at the tip end of the valve stem. This is where a valve "stops" at full-open position, and where the greatest amount of wear occurs.

To check valve stem wear, measure valve stem diameter at the tip end of the valve's stem-to-guide contact surface. For comparison, measure the valve stem diameter between the worn area and the lower valve keeper groove. The difference between these two figures is the amount of wear.

Now, how much wear is too much wear? There are so many variables that there is no one correct answer for all occasions. For instance, what are you going to do with the guides? Bronze

A hand-operated pneumatic motor powers the grinding stones that cut the new valve seat.

guides can tolerate more wear than cast-iron ones. If you choose knurled guides, I suggest there be no valve stem wear. Secondly, what kind of service do you expect from the engine? All of these must be considered together to arrive at a logical decision on the amount of permissible wear.

Chrysler says 0.002-in. is maximum permissible wear; the machine shops use the figure 0.0015-in. The best plan is to use factory specifications for new stem-to-guide clearance: 0.001 to 0.003-in. on the intake and 0.002 to 0.004 on the exhaust. Remember, the 426 has a little more play—0.002 to 0.004-in. for the intake and 0.003 to 0.005-in. for the exhaust. The nearer you approach these clearances, the better service you'll receive from your heads.

Reconditioning the Valves

So, most of your valves check out. The next step is to grind the face to match the seat. This ensures proper sealing. To do this, the machinist uses a special machine for grinding valves. The equipment used is quite sophisticated and requires a bit of skill to use. Therefore, it should be done by the professionals.

The first operation is to resurface the tips. Here, the tip of the valve is ground square to the centerline of the stem. Finally, the tip edge is chamfered (beveled) to remove the sharp edge created while resurfacing the tip.

The valve stem is then inserted in a motor-driven chuck, set at 45° to a grinding wheel. The valve is spun and the grinding wheel passed back and forth across the face. During the process oil bathes the valve face to wash away the grit and cool the valve.

For the face and seat, Chrysler recommends 45°. Some machine shops will grind the face of the valve to 44° instead. The valve face is ground for 1° edge contact at the outermost part of the valve seat. Hot combustion gases will then be sealed the first time the valve closes. Leaks at either valve due to poor seating can warp the valve and burn both seat and face.

Only enough material is removed to restore a void-free 44° valve face. Removing too much material will reduce the valve's margin to the point where the valve must be discarded.

The margin is the width at the outer edge of the valve face. An exhaust margin is exposed to high heat concentration, so when the margin is too narrow, the valve will burn. Valves with margins of less than 0.045 in. should be

Before grinding the face, the tip is ground flat and 90° to the centerline of the valve stem. Then it is chamfered to prevent chipping.

The valve is centered and clamped into a rotating chuck. A rotating grinder then passes across the face of the valve. Oil cools the operation. The valve is finished to an angle of 45°.

This valve spring tester measures free height, solid height, load at installed height and load at open height. It can also serve as an aluminum can crusher!

discarded. Intake valves run cooler so a smaller margin is acceptable.

Once the valves are refaced, you don't need to keep them in order. All traces of mating to other components are gone.

Hand-Lapping Valves—Hand-lapping valves is still used by many of today's do-it-yourself enthusiasts. The process originated many years ago when "shade-tree mechanics" had poor machines at best and no machines at worst, to grind valves. To correct a poor valve job, or often just to do one at the roadside, the mechanic applied an abrasive paste to the valve face and seat. The valve was set into the guide and a *lapping stick*, essentially a suction cup on the end of a dowel, was attached to the head of the valve. The valve was rotated back and forth by rotating the lapping stick with the palms of the hands, much like a Boy Scout trying to start a fire by rotating a long stick into a flat board.

Eventually, enough material was worn away so that the valve and seat had something of matching faces and sealed well enough to get them on down the road a few thousand more miles.

I've mentioned this process because you may have heard about it or your grandfather may have insisted that it was an essential part of the valve job, or you may use it yourself. If it works for you, by all means continue using it. But an accurately done, professional valve job won't need hand-lapping, and all the lapping in the world won't correct a poor job.

Valve Sealing

To insure a perfect valve job, select a machinist with an excellent reputation. To put your mind to rest, however: there is a way to check the valves for sealing after the job is done. Assemble the heads, with or without doing the following valve spring work. Lay the head with the deck surface up. Fill each combustion chamber with kerosene or other solvent and check the ports for any seepage. If none is found, you can be sure the job was well done. If seepage should occur, return the head to the machinist to have it corrected. Now turn your attention to checking the valve springs.

VALVE SPRING INSPECTION

The final phase of inspection and reconditioning of the heads is spring inspection. A valve spring tester is nice but not necessary. Mostly, you should understand the importance of maintaining certain minimum standards and know the terminology. This includes: valve float, spring rate, free height, solid height, load at installed height and load at open height.

Valve Float

Valve springs don't just close valves. After the cam lobe pushed the lifter that pushed the pushrod that pushed the rocker that opened the valve against the force of the spring (wow!) the valve spring returns the compliment and puts all of it back where it came from. The return side of the cam lobe helps by making sure everything is eased into place. But the valve spring must hold the lifter in contact with the lobe on both sides of the lobe.

Valve float, better termed valvetrain float, occurs when engine rpm is so high the springs can't get everything back in time. Lifters momentarily lose contact with the cam lobes and the valvetrain

I like to use a cabinet maker's square to check the valve spring for plumb. The flat base upon which the spring can sit provides excellent accuracy. As you check the spring for plumb, rotate it carefully and slowly. If you simply take one try you can get a "good" reading from a bad spring. This is seen in the photo in the center.

momentarily "floats" away from the cam lobes. Instead of being controlled by the camshaft, the valvetrain is momentarily controlled by inertia, spring rates and a little voodoo called "harmonics."

It all gets a bit too involved to explain fully here but the results are sheer chaos—the valves bounce when they close, they close late, the lifters can't get back where they're supposed to be. The pushrods and rockers are somewhere in between.

To prevent valve float, at least to the desired engine rpm, the valve springs must have sufficient force to overcome the inertia of the valvetrain to make the lifters follow the cam lobe.

Spring Rate

A spring requires a given amount of force to be compressed a given amount, usually expressed as pounds per inch of deflection. Spring rates relate to engine rpm. The higher the rpm, the higher the valve seat load—installed load—must be to control valve float.

Free Height

Free height is the unloaded length of the spring. If free height is too long or too short, the spring will be compressed too much or too little by the spring retainer. The force exerted by the spring when installed—load at installed height—will be off.

Load at Installed Height

This is the load exerted by the spring when compressed to its installed height. Installed height is measured from the cylinder head spring seat to the underside of the spring retainer with the valve closed.

Check each spring individually. Add necessary shims and keep each assembly with its valve.

A typical load at installed height specification might be 121 to 129 lbs. at 1-55/64-in. This means that it should take between 121 and 129 lbs of force to compress a spring to the height of 1-55/64-in.

The absolute minimum limit is 10% less than the minimum standard limit, about 110 lbs. in this example. If a spring does not meet or exceed the minimum it should be replaced.

Load at Open Height

The next specification is spring load when the valve is fully open. The specification might be between 192-208 lbs. at 1-7/16-in. As with the load at the installed height, a spring must fall within the 10% limit or be replaced. In the above example the minimum is 173 lbs.

Solid Height

Solid height occurs when a spring is totally compressed or coil bound. Here, each coil touches the adjacent coils. Valve springs should never be compressed this far in an installation. Obviously, if the spring was totally compressed before maximum valve lift the valve couldn't reach the fully open position. The first time the engine cranked over something would give, probably a pushrod.

Squareness

Squareness, or *plumb,* is simply how straight—square—a valve spring stands on a flat surface. The spring should be concentric with the valve stem. If there is any tilt in the spring, it will side load the valve stem and increase wear between the valve stem and guide. The limit for squareness is 1/16-in. measured from the top of the spring to a vertical surface.

Checking the Springs

Of the characteristics just mentioned, the following should be checked carefully: squareness, installed height and free height. The other three characteristics must be checked with a machine.

Squareness can be checked with a cabinet maker's square and a small machinist's rule. Set the square on the bench with one leg horizontal and the other vertical. Set the spring upright on the horizontal leg. Rotate the spring against the vertical leg until you get the widest gap between the top coil and the square.

With the machinist's rule, measure the gap. If it exceeds 1/16-in. the spring must be replaced. If this all sounds like too much, just think of it as a carpenter measuring the end of a board for square. If the board is more than 1/16-in. off then it must be discarded or finished to remove the 1/16-in. gap.

Another test you can perform is to check spring free height to the specifications in the nearby chart. If a spring has been fatigued or overheated it will sag. By comparing the free height of the old spring to the free height of a new one—or in our case the specified height—you can determine if the old spring is still usable.

To Replace or Not—Generally, a spring more than 1/8-in. under specified free height should be replaced. If, however, you plan any kind of extended use or high rpm, then anything under 1/16-in. should be replaced.

Like piston and valve-stem knurling, there's a "make-do" solution to the short-spring problem. A shim (spacer) can be installed under the spring.

Here again, you get what you pay for. A spring is designed to be compressed a given amount continuously. This amount is the difference between its free height and open height. If it's compressed past its open height the spring will be overstressed, become fatigued and quickly lose its load-producing capacity. If you can possibly afford it, avoid the shim solution and replace under-spec. springs.

Installed Spring Height

A slight but important change has been made to the cylinder head. When the valve seats were ground, material was removed from each seat.

This removal reduced the distance from the valve seat to the spring pad. Consequently, each valve will seat "deeper" into the head and the valve stem—and keeper grooves—will extend farther out of the top of the valve guide.

As a result, valve spring installed height—measured from the spring pad to the bottom of the retainer—has increased. The increase must be corrected by adding shims between the valve spring and its seat or pad.

This is an accepted, safe and necessary practice to correct valve spring installed height. It should not be confused with shimming a weak or partially collapsed spring.

There are two easy methods used to determine installed height. The first is to install the valve into its guide with only the spring retainer and keepers. Omit the spring, seal and any shims.

Seat the valve face in the valve seat by pulling up hard on the retainer, then measure the distance from the spring pad to the bottom of the retainer. This distance can be measured with a

VALVE SPRINGS

Engine	Installed		Open		
Size	Height (in.)	Load (lbs.)	Height	Load	Free Height
383	1 53/64-1 57/64	121-129	1 7/16	192-208	2.58
400	1 53/64-1 57/64	121-129	1 7/16	192-208	2.58
413	1 53/64-1 57/64	121-129	1 7/16	192-208	2.58
426 Hemi					
Outer	1 53/64-1 57/64	102-108	1 13/32	179-189	2.47
Inner	1 41/64	47-53	1 11/64	86-96	2.20
440	1 53/64-1 57/64	100-110	1 23/64	236-256	2.23

This chart gives the measurements of valve springs at installed and open heights and loads.

telescoping gauge and micrometer, or, if you're very careful, with a 6-in. scale.

Another great way to do the job is to cut a heavy wire and file it to the exact installed spring height then check the length with a micrometer. Now, check the spring pad-to-retainer distance and add shims to the pad using the wire as a gauge.

If you used the telescoping gauge or 6-in. rule to check the spring pad-to-retainer distance, subtracting the installed spring height will give you the shim thickness required to correct the installed height. As a formula you have:

Shim Thickness = Spring pad-to-retainer-distance - installed spring height

Shims come in sizes 0.015, 0.030 and 0.060-in. and can be stacked to the required thickness. If the variation is less than 0.015-in., you don't need a shim.

The second way to check spring-to-retainer distance is to install the valve with the springs, retainer and keepers. This is a little more time consuming, but assures a greater degree of accuracy because you're assured that the valve is well seated.

Beyond this, the measuring is the same except you cannot use the telescoping gauge or wire. You can use the 6-in. rule, but the best route is with the depth gauge on a vernier caliper. Whatever method you decide to use, the formula for shim thickness is the same—subtract specified installed spring height from the spring pad-to-retainer distance.

As you check for installed height and determine shim thickness, keep a record of all the parts and their location. Keep the parts organized and matched to each valve as you move along. Once you've checked installed height, the retainers, springs, shims and valves must be installed in the same guide in which they were checked.

With your list in hand, make one more stop at the parts store and purchase the necessary shims. To make your life easier, shims are color coded with the following V.S.I. codes: black = 0.015-in., silver = 0.030-in., and gold = 0.060-in.

ROCKER ARM ASSEMBLIES

Throughout this book we've disassembled, cleaned, inspected reconditioned and reassembled parts. The rocker arm assembly is no different. The most common problem is a restriction within the rocker shaft limiting oil flow to the rocker arms.

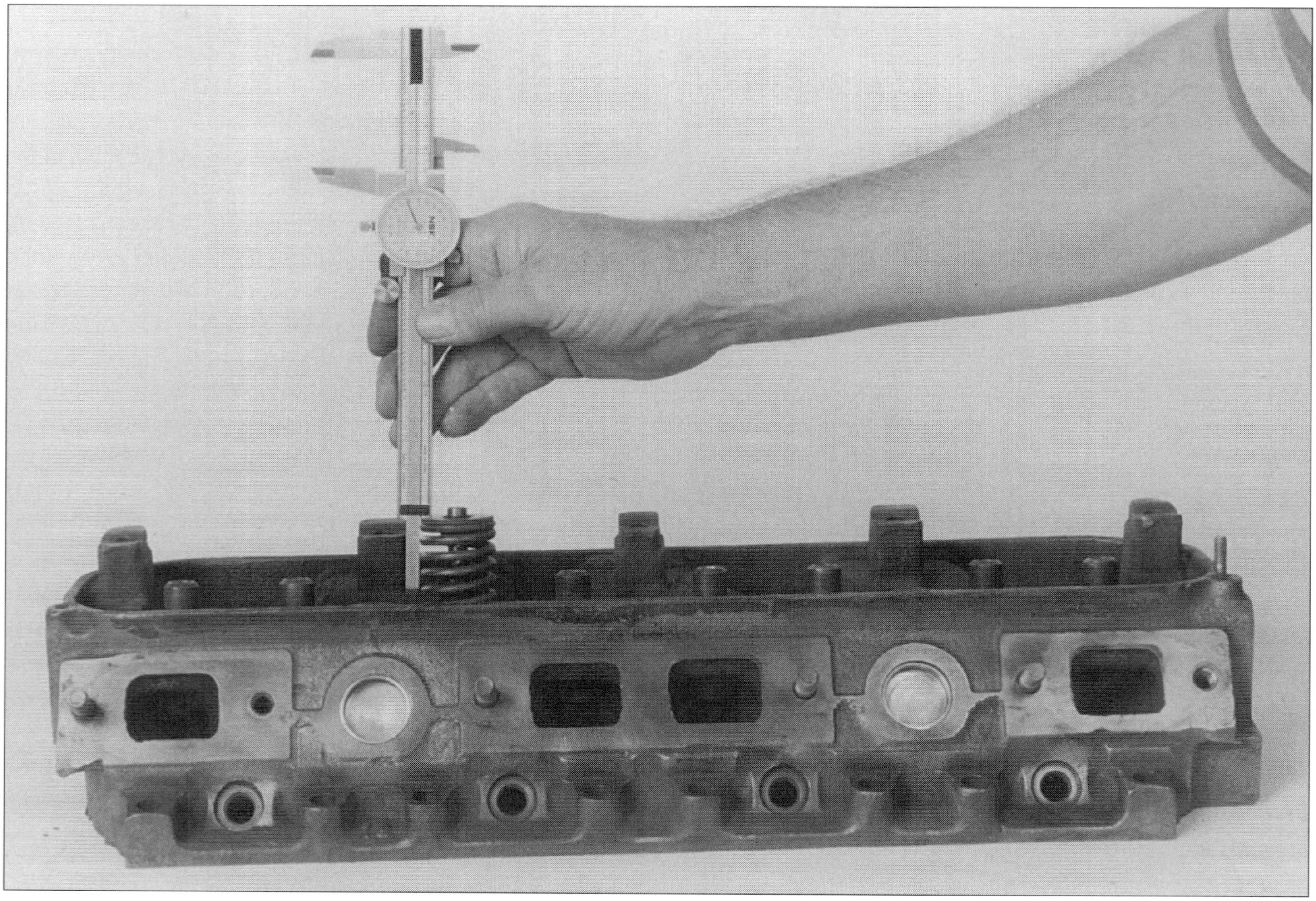

Measure installed spring height from the spring seat to the underside of the spring retainer. For most "B/RB" engines installed spring height is 1-53/64 to 1-57/64 in.

Disassembly

Start by removing the bolts from the rocker arm shaft and slide the rocker arms off. Keep the rocker arms in order. Like the other loaded parts of the engine, each arm has developed a wear pattern on itself and the shaft. Mixing the parts will cause more rapid wear.

Additionally, "B/RB" type engines with hydraulic lifters use offset rockers which means each assembly has left and right rocker arms. Although they appear the same at first glance, the location of the pushrod ball and oiling groove are different. Usually, the rocker arm is stamped "L" or "R." If this is not the case, you'll have to determine left or right for yourself.

Identifying Left or Right—To identify a right or left hydraulic rocker arm, use the following procedure. Compare the location of the pushrod socket. On the left rocker it is located approximately 1/16-in. to the right, and on the right arm, approximately 1/16-in. to the left. Right and left, in this instance does not refer to passenger or driver side, rather to the order in which the rocker arms are arranged on the shaft. Refer to the drawing on page 109 to better understand this.

Before proceeding, check the shaft for obvious damage. If it is badly worn or galled, replace the shaft and save yourself the trouble of cleaning it.

Remove the 5/8-in. plugs from the ends of the shaft. The easiest way is to insert an 1/8-in. punch through the bolt hole nearest the plug. Place it against the back of the plug and drive it out with a small hammer. Don't use the same process you used on the core plugs. This would most likely result in driving it into the shaft, making it impossible to remove. I've had to remove three plugs in more than one rocker arm shaft before.

If you don't like the above method, try this. Drill an 1/8-in. hole in the center of the plug. Use a small slide hammer, the kind you use to pull dents in light sheet metal, with a #8 screw attached. Thread the screw into the hole and pull the plug by sliding the "hammer" back against the handle. Once both plugs are removed you can clean and inspect all the parts.

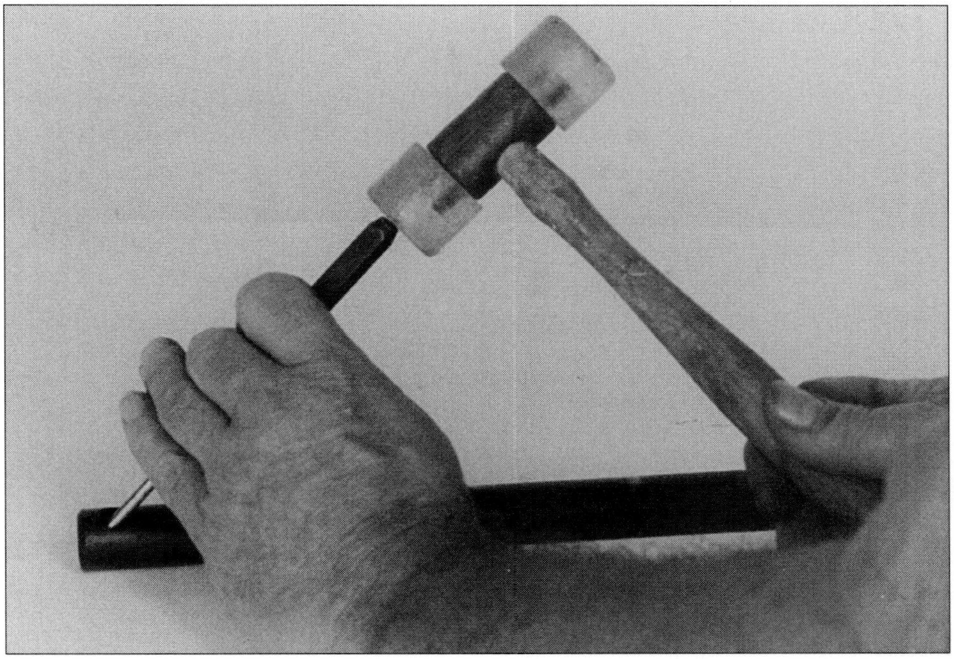

The easiest way to remove the rocker-shaft plug is to drive it out from the back side with a punch.

Cleaning & Inspection

Clean the rocker arms by dipping them in solvent and scrubbing them down with a stiff brush. The rocker arm shaft takes a little more thought. The best tool for cleaning the inside of the shaft is a cleaning brush used to clean the barrels of a 12-gauge shotgun. If you don't have one of these lying around, a trip to the supermarket will get you a baby-bottle brush which will work equally well. Whichever you choose, run the solvent-soaked brush through the shaft a number of times until all oil and sludge is removed.

Again, as you would clean a shotgun, wrap a wire around a 4- or 5-in. square piece of cloth and pull it through the shaft. Do this until a clean cloth remains clean after being passed through the shaft.

Adjustable Rockers—Performance engines are equipped with adjustable rocker arms. These are generally the famous "Max Wedge" heads introduced on the 413 cid in 1962 and on the '63-'64, 426 cid. If you're the proud owner of a 426 Hemi, you'll have double rocker arm shafts with four rocker arms per shaft. It's not impossible, however, to have either a 413 or 426 with non-adjustable rockers. Likewise, there are a few industrial 383's out there with adjustable rockers. Interchangeability plus performance or street use has left us with a lot of "buts" and "howevers." The following inspection procedures apply to adjustable rockers.

Begin your inspection with a close look at the rocker arm shaft. This should be quite smooth with no signs of wear. Minor rust spots or imperfections may be removed with 400-grit emery cloth. A galled or scored shaft must be replaced. If any galling or scoring is found, check the bore of the corresponding rocker arm. It's very likely to be galled or scored also.

Check the clearance between the rocker arms and shaft using a micrometer and telescoping gauge. Measure the diameter of the shaft and the inside bore of the rocker arms. The clearance should be 0.001 to 0.003-in.

If the clearance exceeds 0.003-in., replace the shaft. Adjustable rocker arms can be saved with the aid of the machine shop. The machine shop will bore them, then install bushings and ream them to size. This will give you a better-than-new rocker arm.

Check the rocker arm tips where they work against the valve stem. The easiest way is to simply run your finger over the surface. If you feel any wear, have the tips resurfaced.

Most modern valve-grinding machines can also resurface rocker arm tips. Resurfacing will not only remove the roughness, it will also return the original crown (curve) to the tip. This crown must be perfectly square to the center line of the arm. Do not attempt to do this yourself on the home grinding wheel!

Non-Adjustable Rockers—Non-adjustable, or hydraulic-type rockers, are stamped, rather than cast. For this reason, they must be treated differently.

After measuring the difference between the rocker arm shaft and the rocker arm bore, if you find it to be in excess of 0.003-in., then either the arm or the shaft must be replaced. Usually, the rocker arm is replaced. Unlike its adjustable counterpart, the stamped arm cannot be reworked and used over. Fortunately, they're relatively inexpensive. Replace them if there's any question.

Check the tips for wear. Again, this a touch and feel situation. If any wear can be felt the rocker arm should be replaced. The tips of the arm are case hardened—just like a hammer head. Any wear indicates the hardening has worn through.

Finally, check for cracks. The most susceptible area is around the pivot bore. If you find any cracks, replace the arm.

Rocker Arm Shafts—Assuming wear is not excessive, galling is the major concern. If there are any signs of galling, replace the shaft.

Assembly

Earlier in the book I recommended keeping the rocker arm assemblies together. If you kept them together or kept all the parts in order as you disassembled and checked things, you'll have no problem with the reassembly. If, however, you spilled everything on the floor—or purchased new rocker arms or had the old ones sleeved—then follow the procedure below for the correct method of reassembly. The following directions pertain to all adjustable rocker arm assemblies except the 426 Hemi, which is discussed next.

Adjustable Rocker Arms—Rocker arm shafts are interchangeable from one head to the other. However, each shaft has a 3/16-in. diameter hole which is offset from the centerline of the shaft by 15°. This allows the oil to run forward to the tip of the arm, insuring proper oiling between the rocker arm and valve stem. The centerline referred to can be imagined as the bolt which attaches the shaft assembly to the head. For further clarification, refer to the drawing below.

Set the rocker arm on the head and be sure the oil holes point forward. Now mark these shafts right and left or driver/passenger and carry them over to the bench.

Lay the shaft in front of you so that as you install the rocker arms, the tips—valve end—point away from you. Half of the rocker arms are angled to the right and half to the left. When installed, with the tips pointing away from your body, the arm farthest to your left should angle to the left. The next should angle to the right. Between each left/right pair, install a spacer spring.

When all eight rocker arms are installed, slip a bolt or piece of wire through the bolt holes in the ends to keep the arms from sliding off. Spray a little WD-40 or other penetrating oil on the assembly and store it in a plastic trash bag until you're ready to install it.

426 Hemi—The double rocker arm assembly of the Hemi makes it a bit more complex than the single rocker arm assembly. We'll walk through it though, and everything should work out fine. I suggest you look carefully at the photo and drawing nearby as you work along.

Before we begin, let's discuss the rocker shaft brackets. There are five for

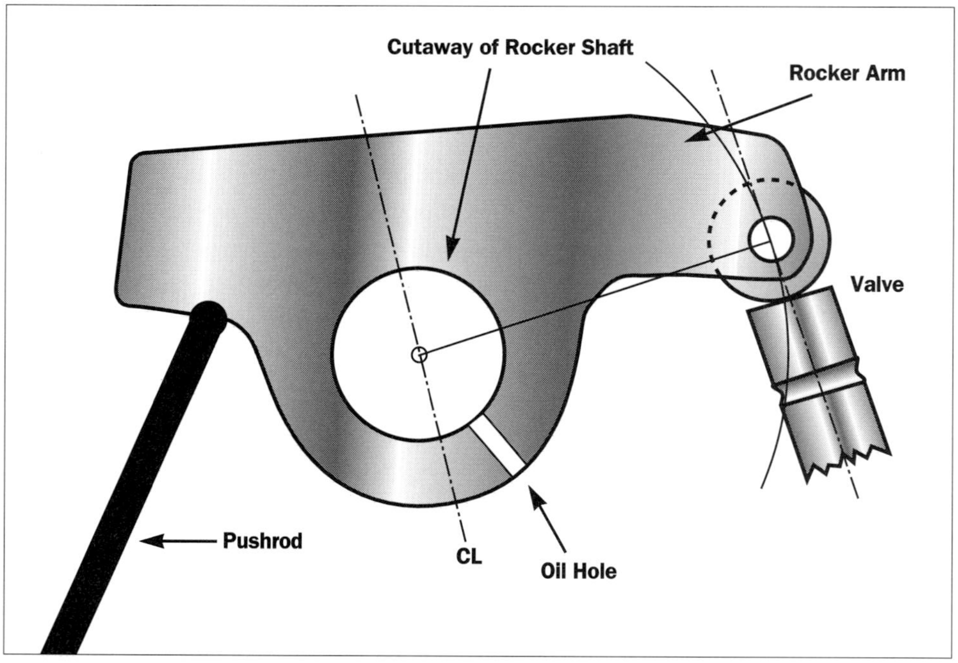

This is a cutaway view of the rocker shaft. Note the location of the oiling hole. This 15° offset from the centerline allows the oil to run forward to the tip of the rocker. Excessive wear will occur at the tip if the oiling hole position is reversed.

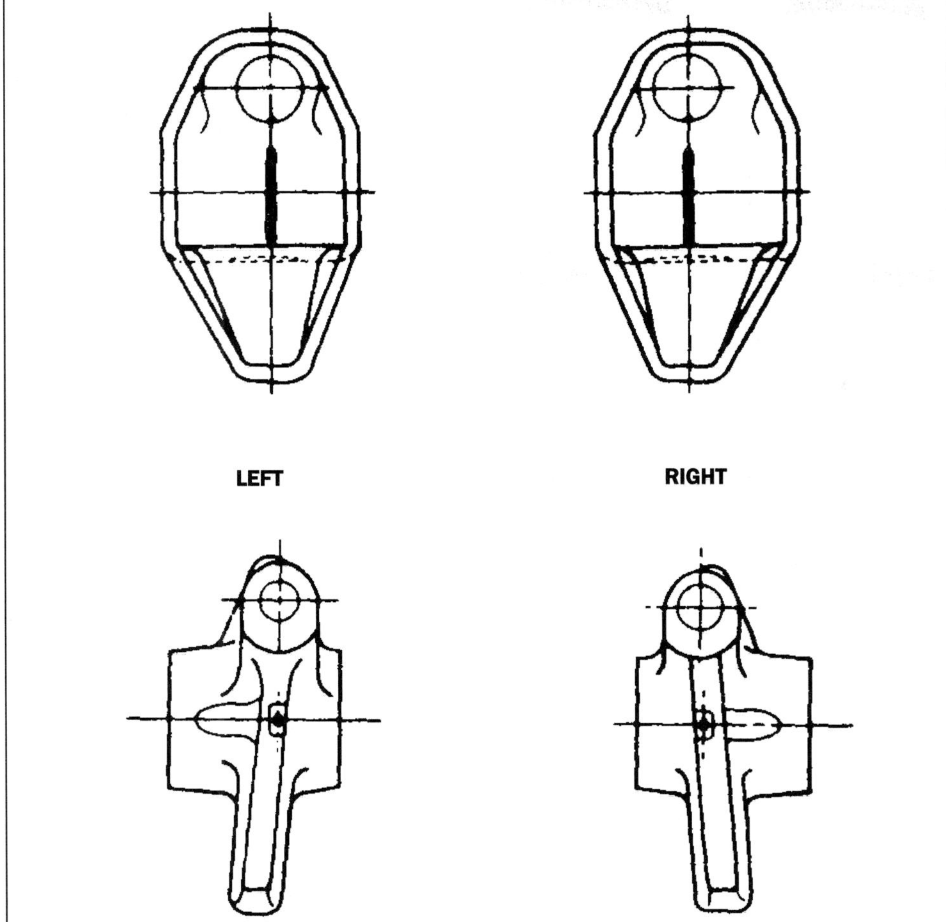

Use this illustration to determine your right and left rocker arms if you were unable to keep them in the order they were removed.

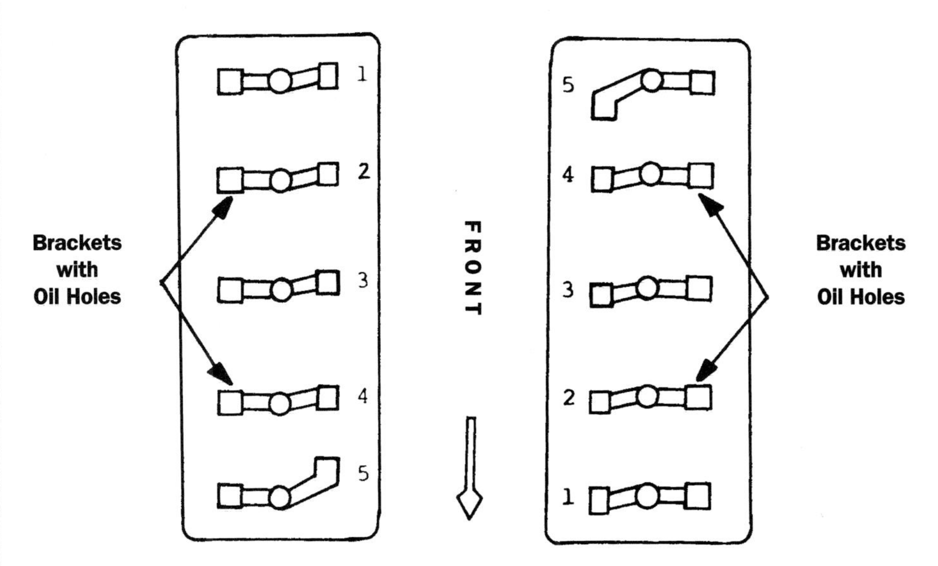

When assembling the rocker arm shafts on the 426 Hemi, be sure that brackets number 2 and 4 are correctly located and are the brackets with the oiling holes.

each side and four of them look the same. However, they're not the same, as two of them are ported to carry oil to the shafts. Look at the drawing below. Note that brackets number 2 and 4 are the two responsible for oiling and have oil holes visible from the bottom. Notice also that both sets of rockers are assembled the same way but reversed in their location on the head. The passenger side is numbered 1 through 5 from rear-to-front and the driver's side is numbered 1 through 5, front-to-rear. Just be sure you assemble the rockers with these two brackets located in the 2 and 4 position. If you fail to do this you'll either destroy the rocker arm assemblies from oil starvation or you'll have a terrible cylinder head gasket leak.

Assemble your parts on the bench with the exhaust shaft (long shaft) closest to your body and the intake shaft (short shaft) on the other side. This would be as if you were assembling the passenger side viewed from the passenger side of the engine. Slide the number 5 bracket—the one heavily angled—onto the two shafts in the position in which they lay.

The short rocker arms are intake rocker arms and the long are exhaust. Slide an exhaust arm over the exhaust shaft with the valve tip pointing toward your body and angled to your left. Slide the short spring over the intake shaft and follow it with an intake arm—the valve tip pointing away from your body and angled to your right. Now slide bracket number 4—with oil hole—onto the shaft. It will angle to the left.

Now install another exhaust arm followed by a spring. Install a spring on the intake shaft and follow it with an intake arm. Slide on number 3 bracket (no oil hole). Repeat this process for the following two exhaust and two intake rockers and brackets. Finish the assembly with bracket number 1. Wire the end brackets—1 and 5—to the shafts, spray the assembly with a penetrating oil and store in a plastic bag until you're ready to install it.

Repeat the above process exactly for the other side (driver). It will be installed in reverse of the passenger side. Now let's move on to something much simpler, non-adjustable rockers.

Non-Adjustable Rockers—Non-adjustable rockers will be found on all big-block Chrysler engines except the Hemi. But as mentioned before, that doesn't mean some big-blocks won't have adjustable rockers. In discussing the adjustable rocker arm assemblies above I noted there was a 3/16-in. oil hole in the rocker shaft. Whether adjustable or non-adjustable, this oil hole, 15° off the center, must point toward the tip of the rocker arm to insure oil flows to the tip. As you assemble the parts, be sure the shaft is in a position such that these holes are pointing away from you toward what would be the outboard side of the engine. If you have any problem with this check the drawing.

The rocker arms are stamped with an L or R. If they are not marked, a close inspection will also show the dimples for the pushrods are offset left or right.

Install the rocker arms on the shaft with the tips pointing away from you in the following sequence: The first arm on the left of the shaft will be marked R. The next will be marked L with each alternating R then L. If you have trouble determining right or left arms, refer to the drawing on the previous page.

When you have both assemblies back together drop in a bolt at each end or run a length of wire through the bolt holes to keep everything in place. Spray with penetrating oil and store in a plastic bag.

CYLINDER HEAD ASSEMBLY

Now comes the fun. Everything is clean and new. The finished product is going to look sharp and you'll have that feeling of accomplishment that comes from knowing it has all been done right and you'll have no worries about the finished product.

Gather your parts together, get out the valve spring compressor, a ruler and possibly a few shims and let's get this all back together.

Core Plugs

You'll need two, 1-1/4-in. brass core plugs for each head. Core plug installation is simple and forthright. Just be careful not to push the plug all the way into the head. Be careful not to bend the edge or push in the center. Either will warp the plug.

Set the plug into the hole, concave side facing out. Try to find a punch, dowel or other instrument that is as close to the inside diameter as possible. It must not be less than 1-in. in diameter. Note in the photo I used the head of a plastic mallet. It fit perfectly.

Tap the plug in carefully and slowly. Be sure it is going in squarely, and don't hit the edges. Stop when the edge of the plug is flush with the edge of the head. Do all four plugs at one time to prevent the chance of forgetting one.

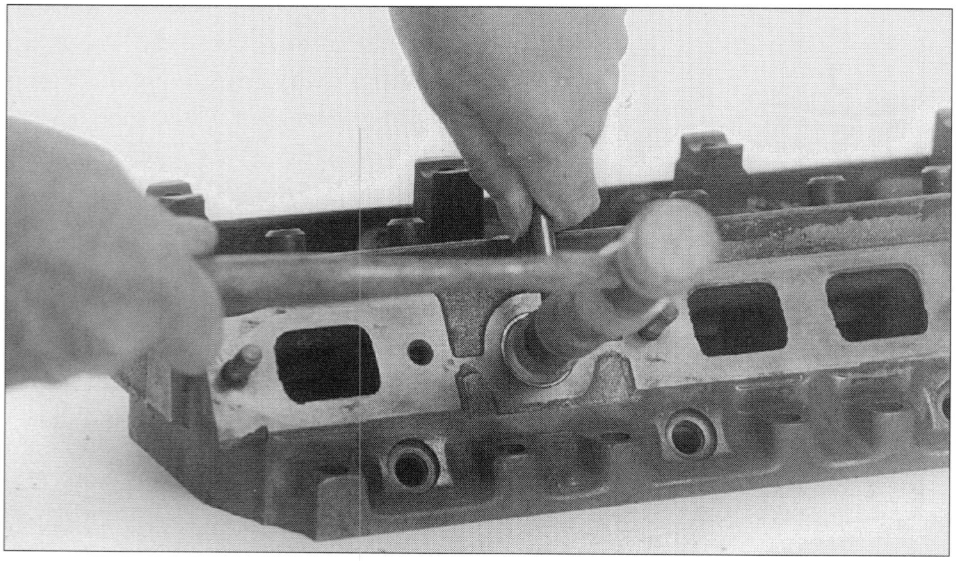

Getting pretty tricky here. I'm using one mallet to drive another mallet to seat the core plug. Note the nice fit of the mallet head, however.

Springs & Valves

Lay the head on its side with the valve seats facing you. Install the valves in the same order you used to measure installed spring height. This will eliminate some confusion. Be sure you know what shims go with each valve and spring.

Place a few drops of oil on the valve stem and a couple of drops in the valve guide. Insert the valve and spin it a couple of times to spread the oil between the stem and guide.

Slide a seal down the stem and seat it firmly against the head. If you're using Teflon seals, the kit should provide a plastic sleeve. This fits over the keeper grooves to prevent them from damaging the seal. If the sleeve is missing, cellophane tape over the grooves and a shot of WD-40 on the seal will work.

Place the shims on the spring pad, set the spring or springs on top of the shims and the retainer on top of the spring. Use the valve spring compressor to compress the spring and retainer to just below the keeper grooves. Do not compress the spring(s) to its solid height. This is hard on the spring and compressor.

Place the two keepers in the grooves and gently release the valve spring compressor. Be sure the keepers are well seated in the grooves and the retainer fits squarely over the keepers.

The rest of the valves are installed the same way. When the last valve is installed, spray the heads with a thin coat of oil, wrap them in plastic and set them aside.

Do not try to drop the valves into the new guides without first oiling the stem and guide. Omitting this step could cause scratches or galling within the guide.

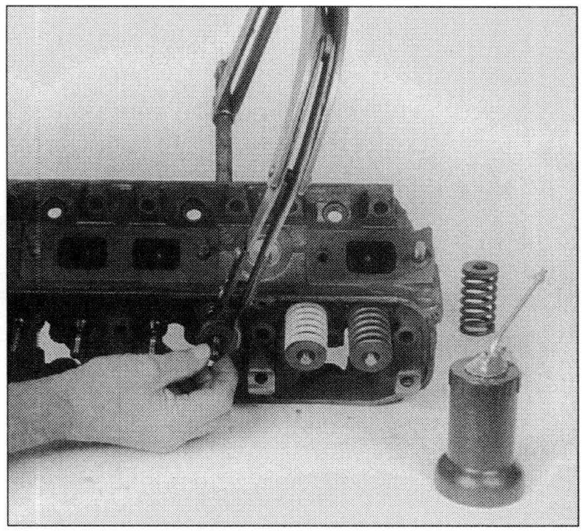

When compressing the valve spring to install the keepers, compress it only enough to expose the keeper grooves, then stop. Compressing the spring until its coils touch is too hard on the spring.

INTAKE MANIFOLD

If you had the intake manifold hot tanked then you've little else to do but inspect it. Look carefully around the carburetor mounting surfaces for cracks. Inspect carefully for any cracks. Most will be visible by eye. If, however, you suspect anything or want to be sure, the intake manifold can be Magnafluxed.

SUMMARY

Congratulations! You're now an official head builder. The important thing in head building is to take no short cuts. If you suspect any problems, correct them. It's so much easier to do it now than later.

In the next chapter we'll assemble the short block and soon you'll be back on the road. ■

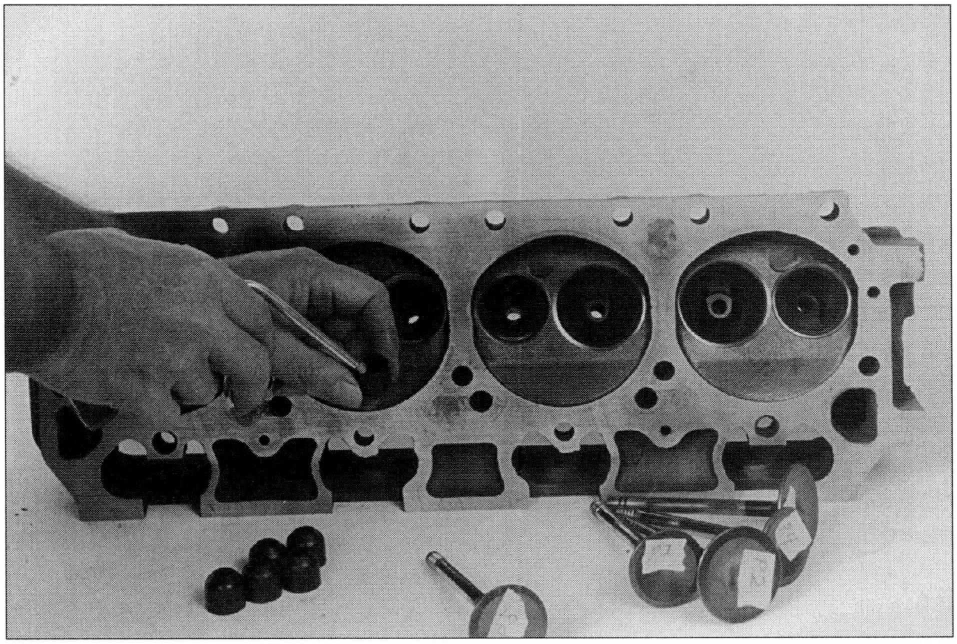

Likewise, a drop of oil on the bore of the seal will prevent tearing it as you insert it over the valve stem.

Here's our completed head. After coating it with oil, I'll bag it up until I'm ready to install it on the block.

INSTALLING VALVE-STEM SEALS

After performing the tests in Chapter 1 you may find the oil-consumption problem to be nothing more than broken or worn valve-stem seals. If that is the case, here's a trick for replacing those worn seals without removing the head from the engine. You'll need two special tools: a spark plug fitting with a male air-hose adapter and a spring compressor. There are several on the market that will compress the valve spring without the head being off the block. Among those are one made by the KD company and the one you see in the photo by Plews Tools (part 72-215).

The trick to the whole process is to pressurize the cylinder. This holds the valve in place while you remove the spring gaining access to the valve seals.

For the demonstration I'm using the assembled engine we've been working with all along. This allows me to photograph the procedure without the distracting elements of the engine compartment in a dark, greasy, environment.

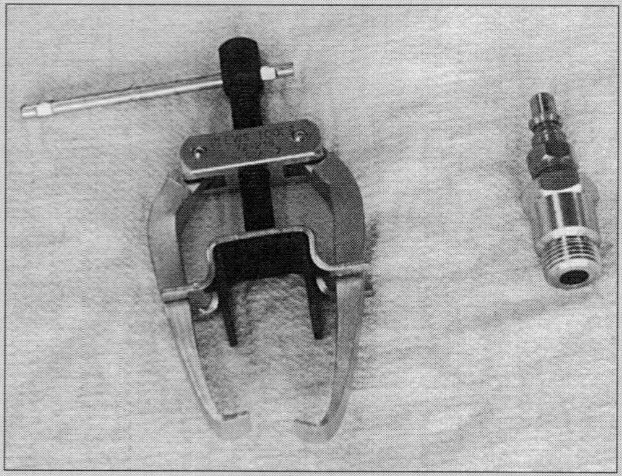

Here are the two tools I'll use. Note that one leg of the spring compressor is longer than the other. This takes into consideration the angle of the spring coil. The adapter to the right threads into the spark plug hole.

Bit by bit, unload the rocker arm assembly. This prevents the possibility of bending the rocker shaft.

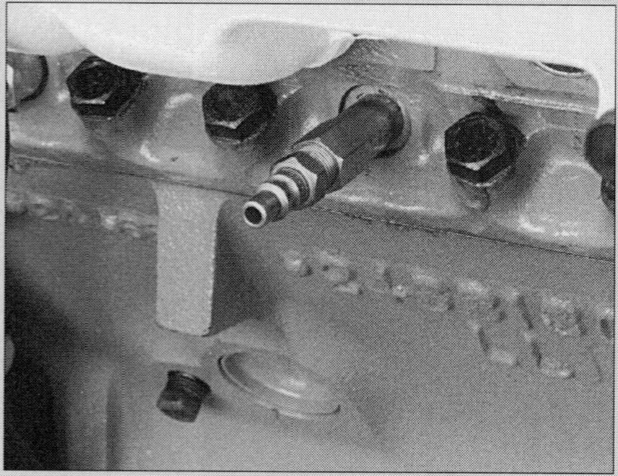

Insert the adapter into the spark plug hole on the cylinder from which you wish to remove the valve spring.

Attach the hose from your compressor to the adapter. You should have the capability of maintaining at least 100 psi.

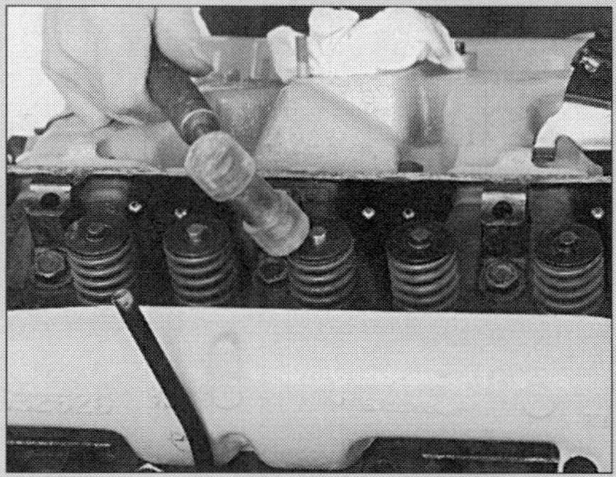

Prior to compressing the spring, give the retainer a few smart raps with your mallet. This will help loosen the keepers.

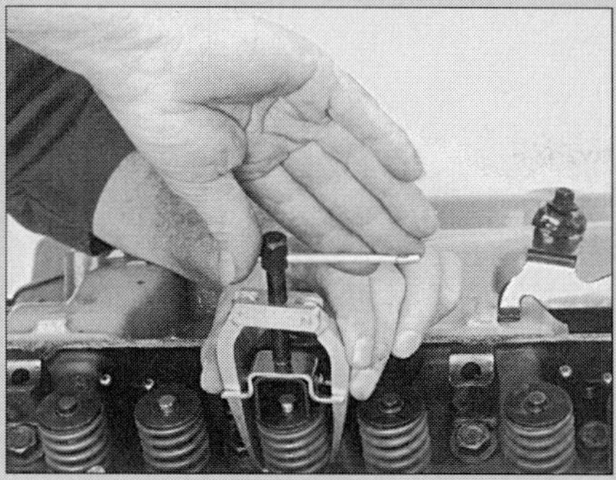

Install the spring compressor per instructions from the manufacturer. If you're using a compressor such as the one in the photo, grab as many coils as you can to compress.

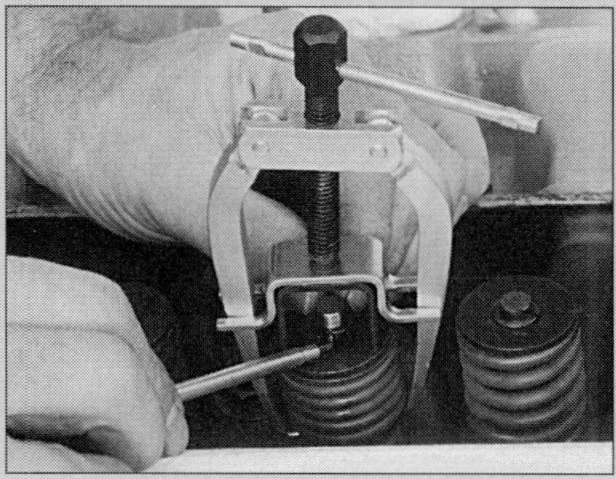

The keepers won't fall out as they did in the demonstration with the head off the block. You'll have to work them out a bit. I'm using a scribe here to demonstrate.

The spring, retainer and keeper are off. What have we here? A broken valve seal. Notice the valve has not fallen into the cylinder. As long as the pressure is maintained, all will be well. If you must leave the valve this way, place a clamp on the stem or wrap tape around it to prevent it from falling in should you begin to lose air pressure.

Included with the valve-stem seals is this little plastic cap. It's installed over the valve stem prior to pushing the seal over the stem. This prevents the keeper grooves from cutting the seal. If you don't find this in your kit, wash the valve stem with lacquer thinner and wrap it with cellophane tape. This will do the same job.

Oil the seal and slide it down over the covered stem. Don't forget to remove the cover!

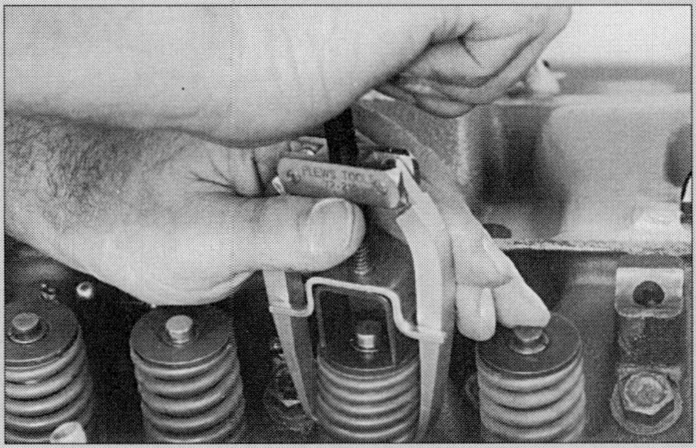

Replace the spring, still compressed. Install the keepers and then release the tension on the spring.

ENGINE ASSEMBLY

7

This is where it all comes together. The really messy stuff is done and only fun is left. You're going to need a few different tools and some materials to aid your assembly. I'll discuss those below. When you get to the actual assembly I'm going to change the format a bit.

Instead of "talking you through" as I'm doing now, I'll create a checklist. As each step is done, check it off. Hopefully, this will prevent you from omitting one or more steps. As you can see, I've tried to illustrate as many steps as possible with photos, but because of a lack of space I wasn't able to illustrate all of them. I suggest you read both the checklist and the numbered photos to make sure you don't miss any steps.

Gaskets

From your favorite parts house buy a complete gasket kit. Fel-Pro and Mr. Gasket are the two top names. I stress complete gasket sets because they're far less expensive than buying just the gaskets you need. A second gasket you'll need that doesn't come with the standard package is the intake manifold/valley pan gasket. This seals the intake manifold to

I thought I'd give you an idea of the end result to get you motivated. You're on the home stretch now. Soon all of those pieces scattered on the bench will be cleanly assembled into a brand-new big-block. Let's get started!

the heads and valley. If you saved your old one and feel it's still usable as a valley pan, you can then buy seals just for the union between manifold and head.

Even though you won't need it now, go ahead and start looking for the carburetor gasket. Most gasket kits that you order will not have one for the carburetor, mainly because there are so many different carburetors to deal with. You'll have to purchase one at your local parts house, and they may have to special-order one. In the town where I live there are six major parts houses. Only one had the gasket I needed and they had to have it delivered from their sister shop in another town. It's been discontinued by Chrysler so don't go looking there either. I mention this now, because you'll need the gasket when you're installing the engine, and it would be frustrating to have to stop in the midst of the installation to search for this gasket. Get it now while you're ordering gaskets to save the hassle later on.

The third "kit" is a set of short-block plugs. This includes all six core plugs, oil gallery plugs, waterjacket plugs, and a cam plug. If you sent your block out to the machine shop they may have replaced these plugs for you.

Sealers

My favorite sealer is clear silicone. It sticks to anything and is the most tenacious sealer on the market. Additionally, it has wonderful gap filling capabilities. But, it's not the only sealer I use. 3M Weatherstrip Adhesive (yellow) is a great contact cement and can be used where extremely fast drying is necessary. This same cement is made by many other brands and is sold as "Yellow Contact Cement."

All of the Permatex sealers are good. They have hard drying, soft drying, high temperature and any number of other types. Follow directions closely (their's and mine) and you'll be very successful with them all.

Additional Materials

You'll need several types of grease and a lot of oil. Buy a medium size tube of moly grease—molybdenum disulfide—a can of anti-seize compound, some old fashioned axle grease and a little petroleum jelly for good measure. The moly grease and anti-seize compound are very important, more so than the axle grease and petroleum jelly. Don't skimp here.

Buy six quarts of oil and don't skimp here either! Your new engine deserves the very best. Why try to save a dollar when you've spent hundreds?

Five quarts of oil go into the crankcase and one quart will be used to oil parts as we go along. You'll need a good squirt can too. Now let's look at some of the tools you'll need.

Tools

The most important specialized tool you'll need is a torque wrench. A torque wrench tells you how much force in ft-lbs. (or metric equivalent) is being applied to the nut or bolt you're tightening. Torque readings are very important when building an engine and must be followed to the number!

There are three kinds of torque wrenches. One has a dial gauge built into the handle so torque can be read while tightening. Another has a screw built into the handle. The amount of torque desired is adjusted on a dial in the handle and the nut or bolt is tightened until the wrench clicks.

The simplest torque wrench has a needle that runs the length of the wrench shaft. The shaft bends when a fastener is being torqued but the needle remains straight. A scale near the handle is calibrated to read in ft-lbs. or in-lbs. You'll need one that reads in ft-lbs. and goes to at least 135 ft-lbs. This is the torque specification for the big crankshaft bolt that retains the harmonic balancer (damper) to the crankshaft.

Next you'll need a ring compressor. This fancy tool compresses the rings around the piston so it will slide easily into its bore. There are two styles. One is like a pair of pliers with a band at the jaws. You tighten the band by squeezing the handle, thereby compressing the rings. The second kind is the band only and is compressed by tightening a secondary band around the first. This is what I use so you'll see it in the photos. The plier style is probably the easiest to use.

Although not really a tool, Plastigage must be described here. Plastigage is a thin strip of soft wax. A piece is placed

1. Buy a full set of gaskets in a kit such as this. Buying one gasket at a time will cost five times as much.

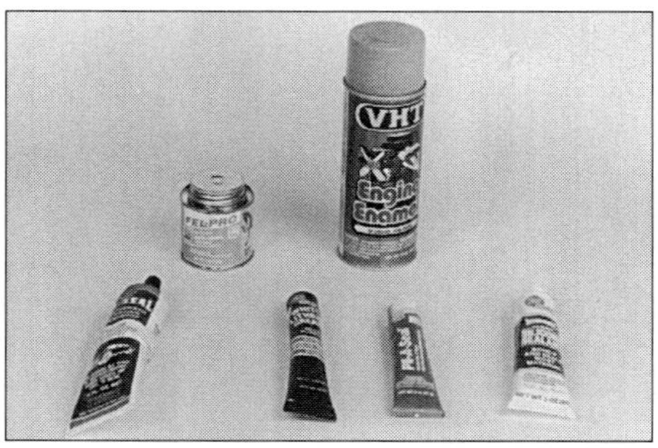

2. Prior to assembling your engine you'll need a number of sealers, greases and some engine paint. The most important are moly grease, anti-seize compound and silicone sealer.

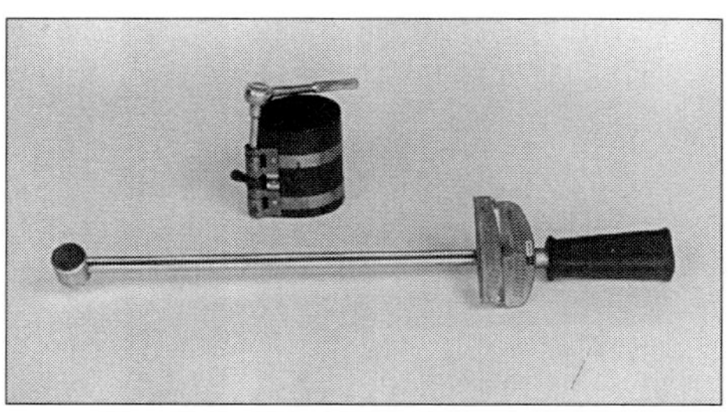

3. Here are two new tools you haven't met yet but will need. At the top is one style of ring compressor and at the bottom is a needle-and-bar torque wrench.

MAKE YOUR OWN GASKETS

As is sometimes the case, one of the gaskets was missing from my gasket kit. It happened to be the oil-pump-to-block gasket. It would have been a simple matter to go buy one but I elected instead to make one so you could see how it's done. I learned this trick watching my father repair the family's 1936 Pontiac sedan by the side of the road during the Depression and in the following years of WW II. This trick goes back a long way.

You'll need some gasket material, scissors or a knife to cut off a piece, a plastic hammer and a small ballpeen hammer. I was able to find gasket material at my local Pep Boys store. Pep Boys is a nationwide chain, but if you don't have one in your area, find the oldest line parts house in your area to locate the gasket material you'll need.

Cut a piece about 2-in. larger than the piece for which you wish to make your gasket. For the oil pump I then cut a hole in the center to fit over the drive.

With your plastic mallet, tap along the edge of the part through the gasket material. The sharp machined edge below will cut through the gasket material, making a perfectly matched gasket.

Use the ball end of the ballpeen hammer to cut the areas around the bolt holes. This must be done very carefully. In fact, I don't cut all the way through. I just mark it and then finish the cut with my shop knife as seen in the photo. In the last photo you can see my perfectly formed gasket.

Sometimes this will save your neck in an emergency. More than once the parts house did not have a gasket I needed. Now, if you find yourself in the same bind you can make your repairs and be on your way—just like Dad and I did as we putted along between Los Angeles and San Diego in 1940.

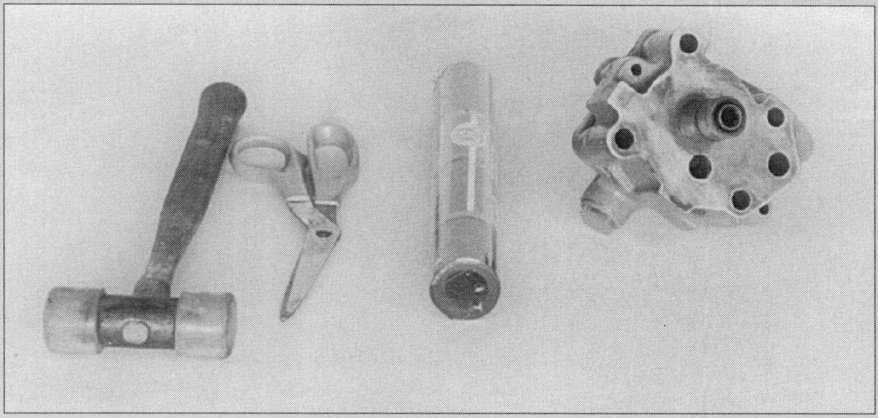

These are the tools of the gasket maker's trade. You'll need gasket material, a pair of scissors or a knife, a mallet and perhaps a ballpeen hammer.

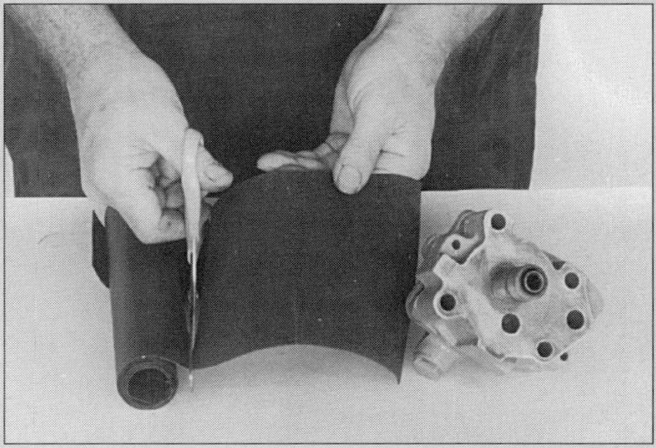

Cut the gasket material oversize to give yourself plenty of working room.

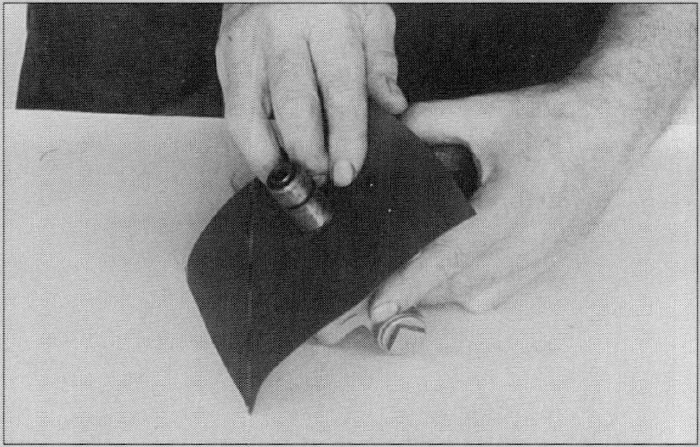

For this project I've cut a hole for the oil-pump pickup shaft.

continued on next page

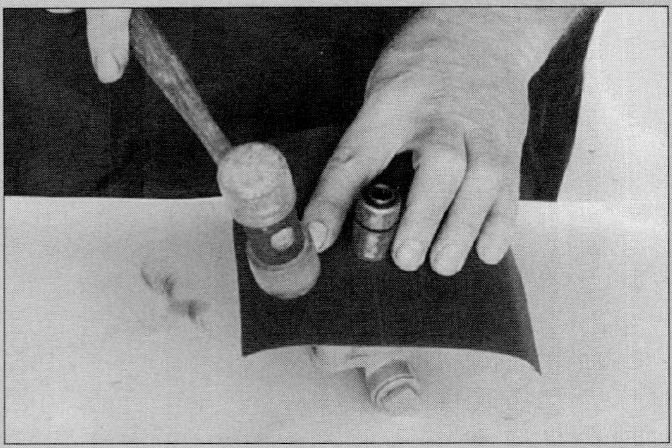
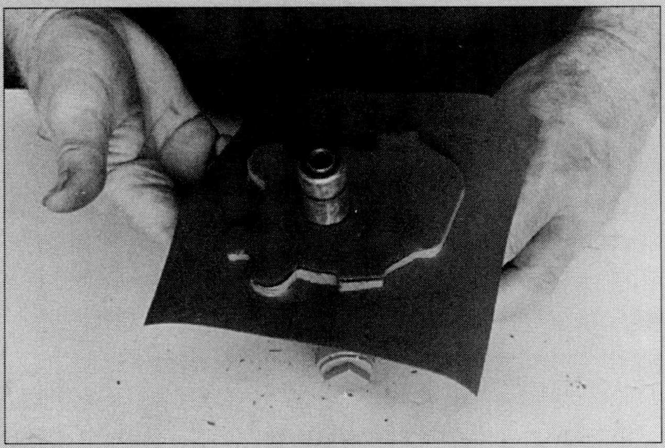

Here's the neat part: the mallet striking the machined metal surface through the gasket material literally cuts the gasket material, leaving a perfectly formed gasket exterior.

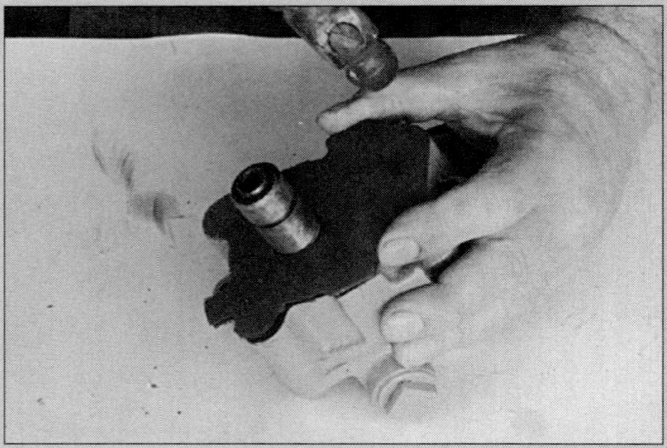

Using a ballpeen hammer on the interior surfaces (around the bolt holes) does the same for the inside parts of our gasket.

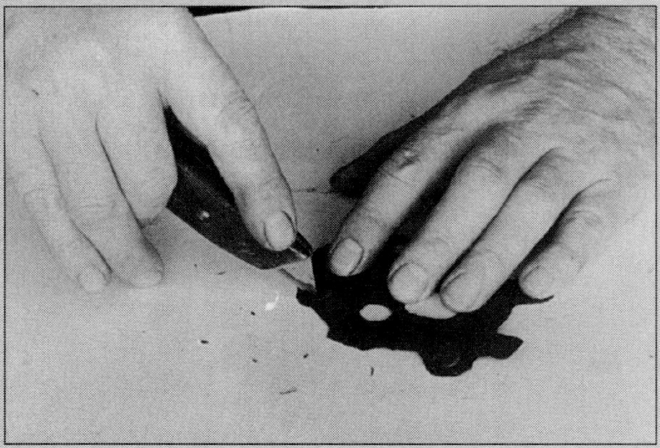

Sometimes you need to do a little trimming with the knife to refine your cut.

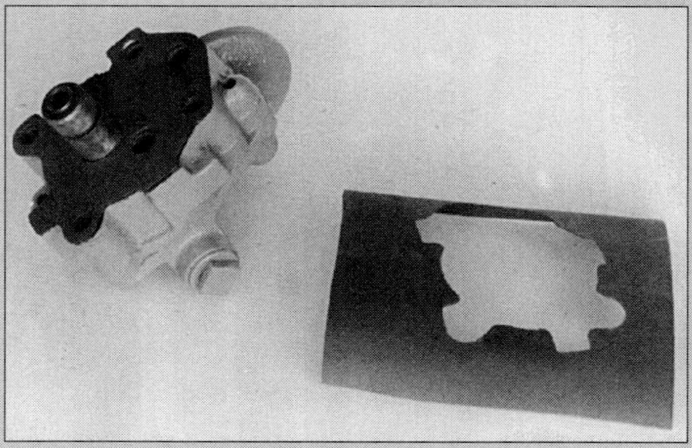

The completed gasket, ready to go. The factory couldn't have done a better job!

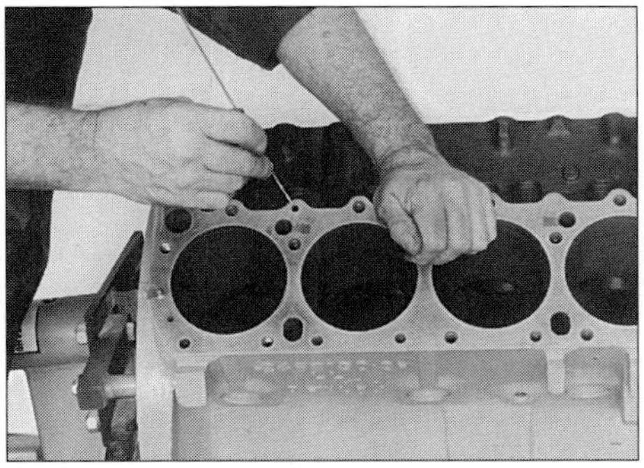

4. This is the oiling hole to the cam bearings. I've inserted a coat hanger wire to be sure the passage is open and the cam bearing hole is aligned with the oil passage. If new bearings were installed without this alignment, your cam bearings will starve for oil.

5. Prior to installing the cam, oil all of the cam bearings. Be sure the entire surface of the bearing is coated. Use your finger for this job.

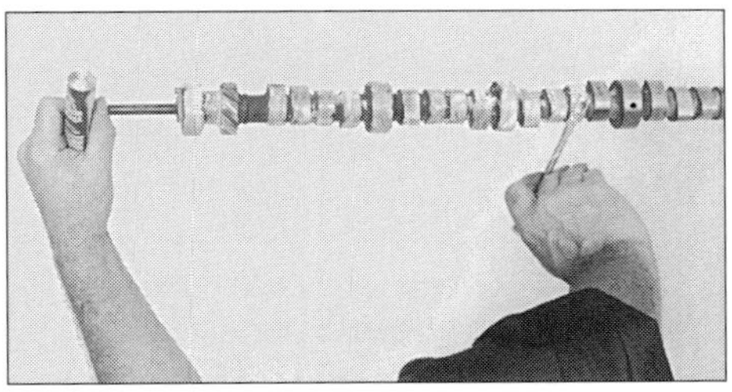

6. Lubricate each lifter and bearing surface with moly grease. This is a must! It takes a while on initial start-up for the bearing to receive oil. This heavy lubrication prevents any possible scoring due to lack of oil in the critical start-up period.

on the journal surface and the main bearing or rod cap tightened to specification. This mashes the wax flat.

The main bearing or rod cap is then removed and the width of the wax measured with the scale provided with the Plastigage. Bearing clearance is determined to 0.001-in. by a conversion scale. Purchase the green—0.001 to 0.003—Plastigage. One strip will be plenty.

7. Okay, here we go. Two hands—one to push and one to guide. Be very, very careful not to bang into a bearing. If the cam binds when rotated, the bearings may be out of alignment. That means taking the block back to the machine shop and having the bearings "blued and scraped."

PREPARING THE BLOCK

You now have all your parts together to assemble the engine. Clear off and sterilize your workbench, give yourself plenty of room to work around the engine and let's put it together.

❑ Remove the plastic wrap from your clean engine block.

❑ If not already mounted to the engine stand, mount it now.

❑ Rotate block into the upright position: valley up.

❑ Check all core plugs, oil gallery plugs and waterjacket plugs—1 on each side of the block.

❑ If core plugs were not installed at the machine shop, coat them with silicone sealer and install as per described on page 110.

❑ Be sure the cam plug is installed. Cam plug is located at the rear of the block.

❑ Do not install the fuel pump actuator rod access plug at this time. This plug is found just below the mounting surface of the fuel pump.

CAMSHAFT INSTALLATION

❑ Rotate the block into the upside-down position. Main bearing webs are in the up position now.

❑ Look at photo #4 and check alignment of oil passages and cam bearings. There is one oil passage on each side.

❑ With your oil can, oil each cam bearing. Be sure bearing surface is completely coated.

❑ Coat all cam lobes and cam bearing surfaces generously with moly (or white lithium) grease.

❑ Install a bolt into the front of the camshaft to use as a handle.

❑ Insert the cam. Use one hand to insert and the other to guide. Pay close attention not to nick the bearings.

❑ Cam should rotate freely by hand. If cam does not rotate freely (any binding at all) cam bearings must be "blued and scraped" at the machine shop. Do not force the cam with a wrench or pliers

❑ Install fuel pump actuator rod and secure by installing the access plug.

CRANKSHAFT INSTALLATION

❑ Wash crankshaft thoroughly with lacquer thinner, MEK or Toluene. Do not use an oil-based solvent such as paint thinner or parts cleaning solvent. The oil will distort the Plastigage reading.

❑ Install main bearing inserts. Top half is grooved and goes into the block. Do not force or hammer the bearing insert. If extra force is needed, gently push on the end of the bearing with the face of a plastic hammer or the end of a hardwood dowel.

❑ Do not install rear main seal at this time.

❑ Remove burrs and nicks from main bearing caps by carefully sanding or filing.

❑ Carefully lay the crank into the block with no grease on the bearings. Do not rotate the crank at any time during the following operation lest you score a bearing or journal. You are only checking clearence now.

❑ Cut off a 3/4-in. piece of Plastigage and lay it across the crank journal you wish to check.

❑ Without moving the piece of Plastigage, install the cap (with bearing) and torque bolts to 85 ft-lbs.

❑ Remove the cap and measure the flattened Plastigage with the scale on the package. If the clearance measures less

8. In it goes. Note the bolt that I used as a handle.

9. The fuel pump actuator rod will serve to hold the cam in position during the rest of the assembly. Install it now.

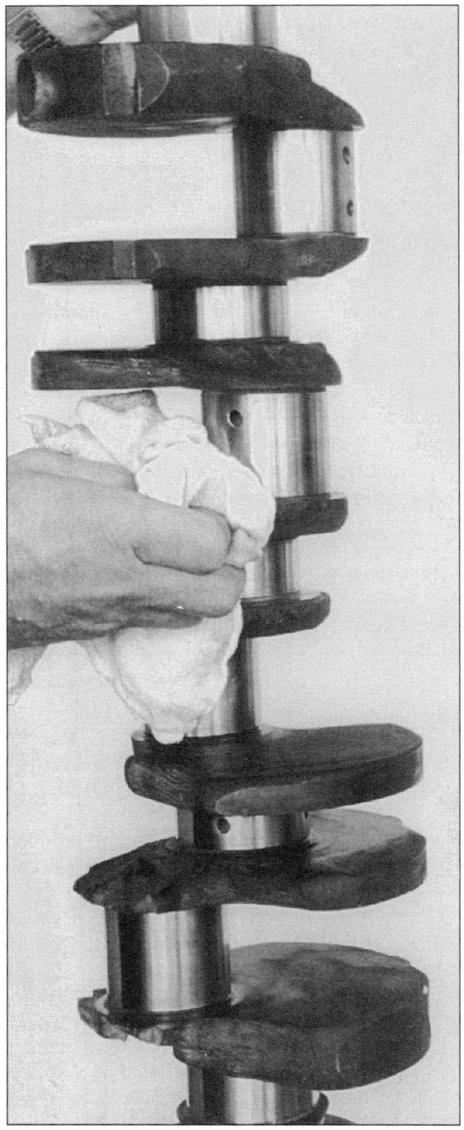

11. Clean the crank with a non-oil base cleaner such as lacquer thinner prior to measuring clearances with Plastigage. Oil-base cleaners will smear the Plastigage.

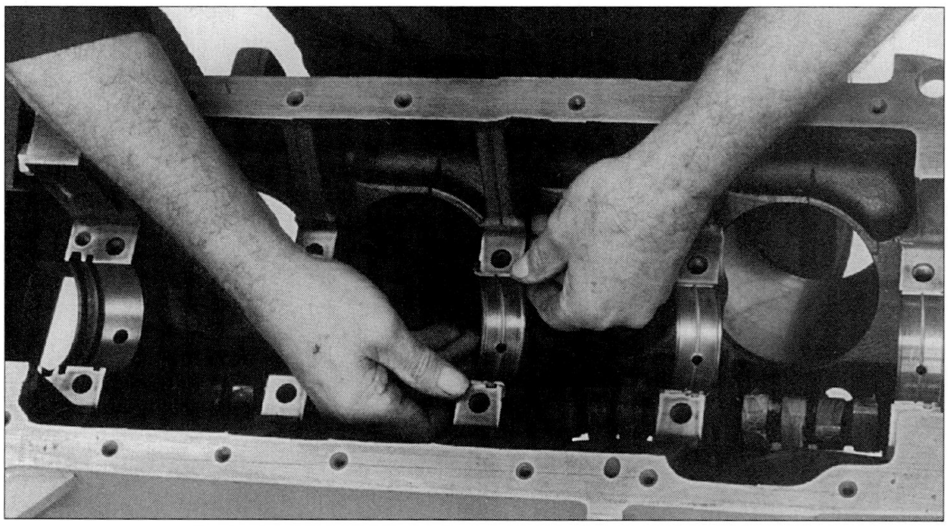

10. Install the bottom half of the crank bearing. Note the oil groove. The top bearing half does not have this groove. Use only finger pressure to seat the bearing. If further pressure is needed, push with the wooden end of your hammer on the end of the bearing only. Never press in the center.

than 0.001-in. or more than 0.003-in. there's a problem. Remove the crank, recheck bearings and journals.

Rear Main Seal

❑ Whenever possible, select a neoprene split-lip seal over a rope seal.

❑ Install the bottom half of the seal in the rearmost groove of the crankshaft bed. The lip of the neoprene seal points forward, ends are flush with the block. A little oil helps work it in.

❑ Rope seals go in edgewise. Push the seal in with your thumb leaving a little sticking above the block. Work the seal in with a few gentle raps of a hammer on the side of a 1-1/2-in. socket. When fully seated, trim ends flush to the block with a single-edge razor.

❑ Prepare the rear bearing seal retainer in the same way. This retainer has seals along the side also. Cement these two seals into the retainer with yellow contact adhesive according to the directions on the label. The edges with the greatest bevel point forward. Set aside and allow to dry during installation of the crank.

Laying the Crankshaft

❑ Be sure everything is scrupulously clean.

❑ Apply a heavy coat of moly grease to all bearing surfaces.

❑ Apply a few drops of oil to mix with the moly grease.

❑ Gently lay the crank onto the bearing surfaces.

❑ Install the five main bearing caps that were pre-lubricated. It may take a few raps with a plastic mallet to seat them. Do not force them. Be sure they are in the same order and direction as originally installed.

❑ Coat the threads and cap-to-bolt mating surfaces with anti-seize compound. This will insure accurate torquing.

❑ Hand-tighten all bolts.

❑ Snug up all bolts with a ratchet and socket. This will insure the caps are well seated.

❑ Torque the bolts in caps 1, 2, 4 and 5 to 85 ft-lbs. in the following manner: tighten one side of the cap to 25 or 30 ft-lbs., then do the same to the other side. Go back and tighten the first bolt to 50 or 60 ft-lbs. and repeat on the other side. This alternating prevents undue stress on the bearing cap that would otherwise occur if you tightened one bolt at a time to its full torque. Finally, tighten each bolt to its full 85 ft-lbs., 100 ft-lbs. for the 426 Hemi. Install the Hemi main bearing cross-bolts (tie bolts) and torque to 45 ft-lbs.

❑ With a couple of big screwdrivers, gently and carefully pry back and forth between two of the bearings caps—not the center one—and crankshaft throws. This lines up the bearing thrust faces. Do this a couple of times then pry the crankshaft forward. Hold it in this position while you torque the two bolts securing the center, or number 3, cap.

❑ Install crankshaft rear bearing seal retainer after carefully oiling the lip seal. Do not get oil on any of the metal mating surfaces. Apply a medium thin coat of silicone sealer to the sides and bottom of the seal/retainer. Push into place and torque the two retainer bolts to 30 ft-lbs. These bolts require the use of a 3/8-in., 12-point socket.

❑ Measure crankshaft end play. Pry the crank as far forward as it will go.

12. After setting the crank onto the bearings do not move it. If you do, you'll scratch the unlubricated bearings.

13. The worm-like line on the surface of the crank journal is my piece of Plastigage. Try to keep it as straight as possible for the most accurate measurement. I made mine a little "wiggly" so you could better recognize it in the photo.

14. Torque the bearing cap to 85 ft-lbs.

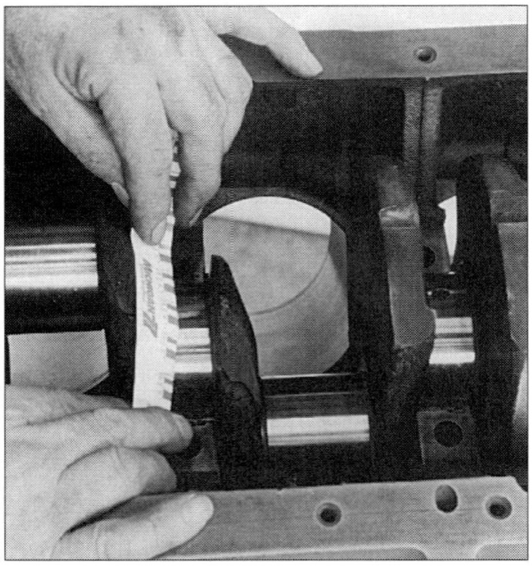

15. Use the scale on the wrapper to measure the flattened Plastigage. Now you know how much clearance there is between the journal and bearing. It's a good idea to test all the bearing/journal combinations—even the connecting rods. After checking, remove crank.

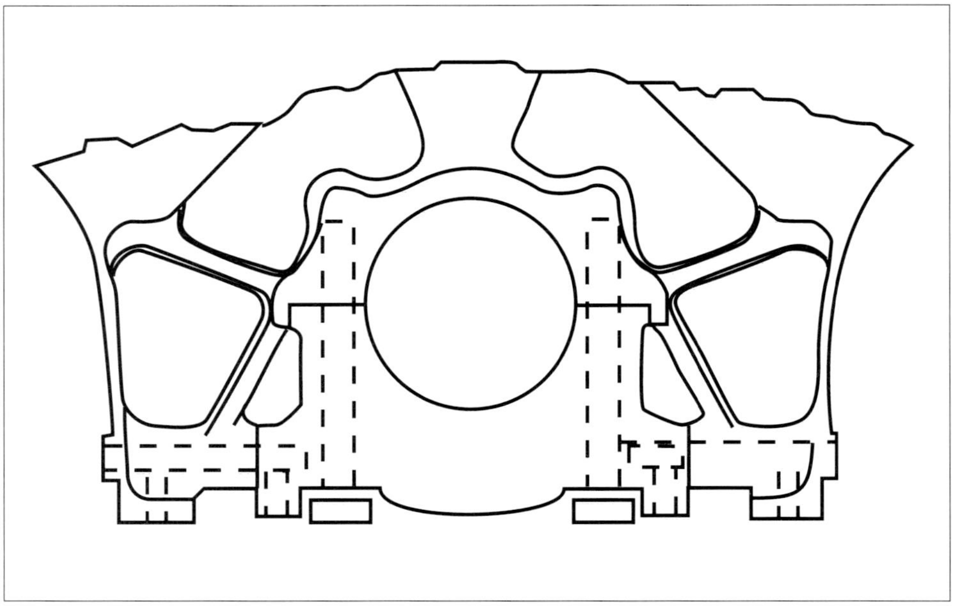

The 426 Hemi main bearing caps have four bolts. Study this drawing for the correct location of the horizontal tie-bolts.

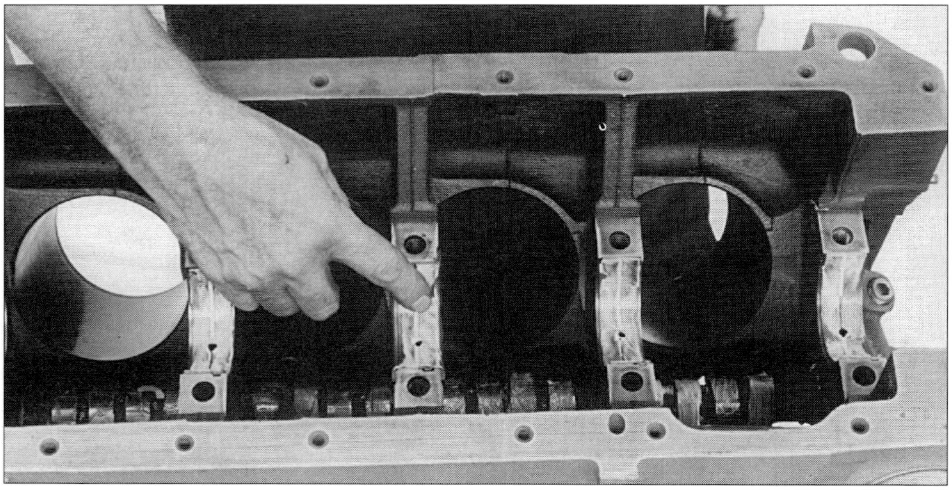

16. As with the cam, lubricate all the crank bearings with moly grease for final installation. For further protection add a few drops of oil, then carefully lay the crank.

17. Here's your first use of anti-seize compound. Coat the threads and the bearing cap surface below the bolt head. This will prevent any seizing of the bolt as you torque it.

18. First, work all the caps down a little at a time with your ratchet. Then torque each bolt a little at a time. Don't try to bring the torque up to full measure on any one bolt until the others are at about the same ft-lbs. Leave the number 3 cap for last. Torque all caps to 85 ft-lbs.

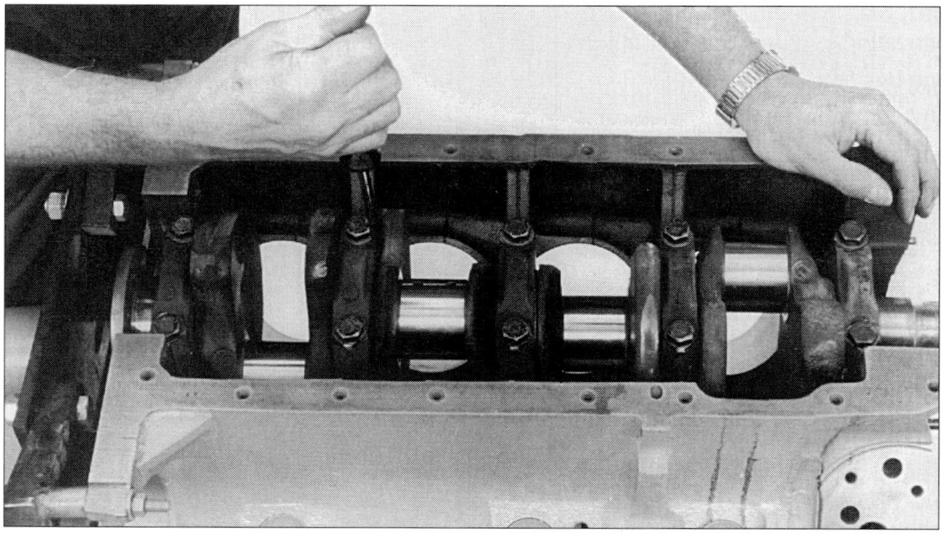

19. To align the thrust bearing, move the crank back and forth with a large screwdriver. Pry the crank forward and hold it while you torque the number 3 cap bolts.

20. The crankshaft rear bearing retainer must be sealed along the sides. Be sure the two gaskets on the sides have been sealed in place. Now smear a thin coat of silicone sealer on the sides and install. Check the neoprene seal to be sure the lip points forward.

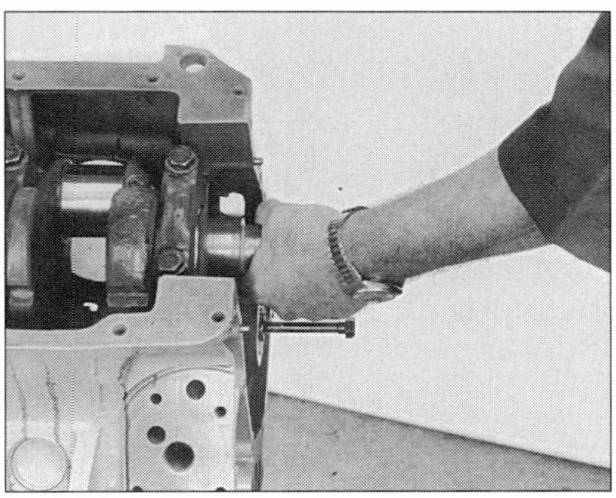

21. If you used the neoprene rear seal, you should now be able to turn the crank by hand.

22. If you used a rope seal, breakaway torque should be no greater than 15 ft-lbs.

Insert a feeler gauge between the thrust bearing flange and the crankshaft thrust surface. The range should be between 0.002 and 0.007-in. In no case should it exceed 0.010-in.

❑ Check crank rotation. If you used a lip seal the crank should rotate freely and easily by hand. If you used a rope seal, breakaway torque should not exceed 15 ft-lbs.

TIMING CHAIN & SPROCKETS

❑ Be sure the block is in the upside-down position: crankshaft up, camshaft below.

❑ Replace timing chain and sprocket if wear so indicates.

❑ Be sure the Woodruff key is in place in the crankshaft and level with the damper shaft.

❑ Install crank sprocket, aligning Woodruff key and sprocket keyway.

❑ Install cam sprocket. Locating dowel on the cam inserts into the locating hole in the sprocket.

❑ The front of each sprocket has an "o" mark stamped into the front. Be sure you can see these marks and that they are indeed pointing to the front. Some aftermarket cam sprockets use a "◊" or a "/" symbol.

Cam & Crank Timing

❑ Rotate the crank and cam until these two marks are as close together as possible. The mark on the crank sprocket will be straight down and the mark on the cam will be straight up.

❑ Using a machinist's straight-edge against the cam sprocket on center with the cam and the crankshaft, rotate both cam and crank sprockets until both timing marks are aligned. When the two timing marks are aligned with the centers of the cam and crank sprockets, crank and cam are correctly timed.

Install Timing Chain

❑ Remove the cam sprocket. Be extremely careful not to move the cam or crank.

❑ Slide the crank sprocket to the end of the crank, but don't remove it.

❑ Drape the timing chain over the crank sprocket and lay the cam sprocket into the loop with the timing marks aligned.

❑ Slide the cam gear onto the cam and simultaneously push the crank sprocket into place. A little wiggling helps slide the crank sprocket over the Woodruff key.

❑ Recheck your work. If off, make

corrections until the timing marks are aligned and on a centerline between the cam and crank.

❑ Install the camshaft lockbolt and washer. Tighten to 35 ft-lbs.

❑ Finish this portion of the job by installing the oil slinger over the crank sprocket.

PISTON RING INSTALLATION

Have the following materials and tools on hand for the job: a small flat file, ring compressor and a ring expander.

Checking Ring End Gap

❑ Install a compression ring in the bore.

❑ Push it in about 1/2-in. with the top of a piston to square it.

❑ Measure the gap between the two ends with a feeler gauge. Compression rings should have a range of 0.013 to 0.025-in. The oil ring side rails are much wider at 0.015 to 0.062-in.

❑ If ring gap is too small, file some material off one end of the ring.

❑ To file a ring, mount the file in a vice and stroke the ring across the file. Stroke from the outside edge toward the inside edge. On moly rings the coating is on the outside edge. The above method will prevent chipping or peeling away the moly coating.

❑ Remember to check the entire set of rings: 8-#1 compression, 8-#2 compression and 16-#3 oil side rails.

❑ Be sure not to mix up the #1 and #2 compression rings. Each is different in its construction. For even more accurate building, measure ring gap by

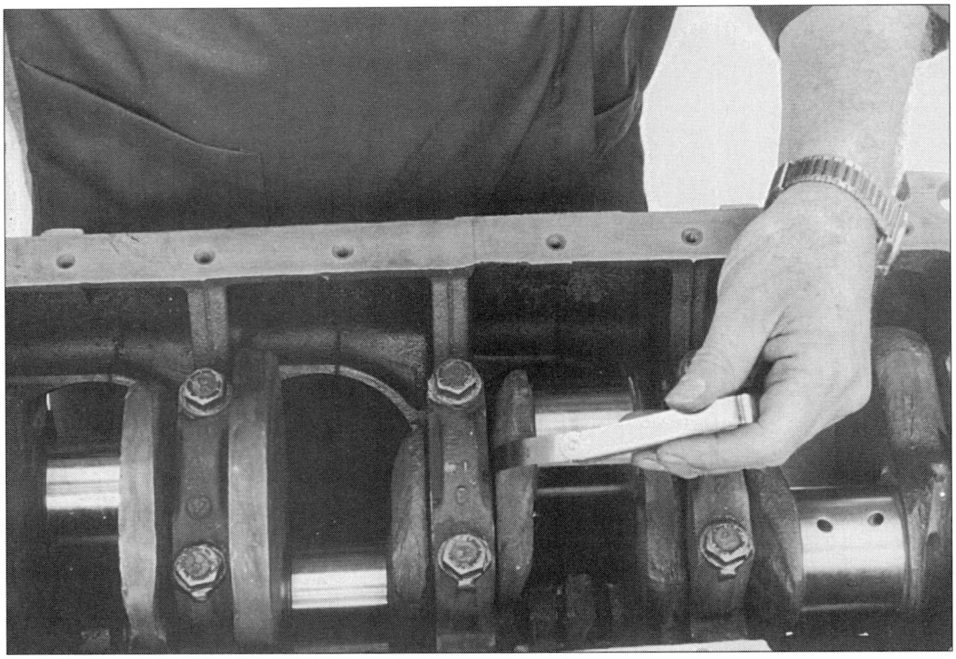

23. Pry the crank forward with a screwdriver and measure crankshaft end play with a feeler gauge. End play should be 0.002-0.007-in., with an absolute max of 0.010-in.

24. On the left is an aftermarket cam sprocket. On the right is the OEM sprocket I removed from the engine. Note the broken plastic covering over the teeth. Had I installed the old sprocket, I'd be replacing it in less than a couple of thousand miles. Do it now while it's so accessible.

25. To correctly time the crankshaft and cam together, the timing marks on each must face one another on a centerline between crank and cam. I'm using a machinist's rule to be sure everything is aligned.

26. After torquing the cam bolt, spin the cam and crank through a couple of revolutions just to be sure everything is working together.

27. Don't forget the oil slinger. The timing chain case cover goes on later.

set and by cylinder. Install the rings on the piston as they're measured and fitted.

Install the Rings

When you install your rings, always follow the directions for placement you find on the package, even if they differ from mine. Each manufacturer makes several different styles and each has its own installation method.

For under $10.00 you can buy a ring expander. This is a great help and is worth every penny you pay for it even if you throw it away after the job is finished! You can do it by hand but you're guaranteed a couple of finger cuts and a scratch or two on the piston—not to mention the real possibility of breaking a ring!

Oil Rings First

❑ Install expander spacer first. Ends must butt together. They do not overlap.

❑ Install side rails. I start with the bottom rail then install the top. You may do it either way you find satisfactory.

❑ Locate the gaps as follows: expander-spacer gap facing the front of the engine with one side rail 90° to the right and the other 90° to the left.

#2 Compression Ring Next

❑ Check directions on the ring package for "which way is up." There is a right side "up" and it's important that you install the ring in the fashion the directions indicate.

❑ If you have a ring expander, insert one end in each side of the jaws, expand the ring and slide it down over the piston.

❑ If installing by hand, mount the piston in a vice, clamping the rod gently—with just enough pressure to hold it still.

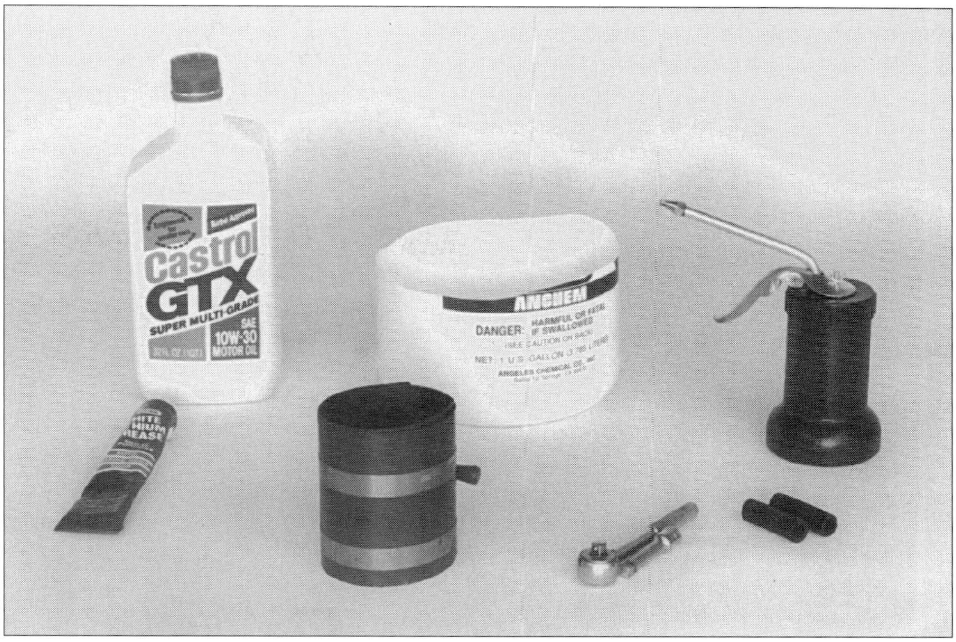

28. Here's everything you need to install the pistons. The 1/4-in. ratchet operates the ring compressor. The two sections of 3/8-in. ID hose serve as rod-bolt sleeves. It's easier for me to cut off a couple lengths of hose than to keep track of my fancy little plastic covers.

29. When measuring ring end gap, be sure the ring is straight in the bore by pushing it in a bit with the top of a piston.

30. Out comes the feeler gauge again. Measure the end gap. If it's too great, exchange the rings for the correct size. If the gap is too small, file a bit off the ends.

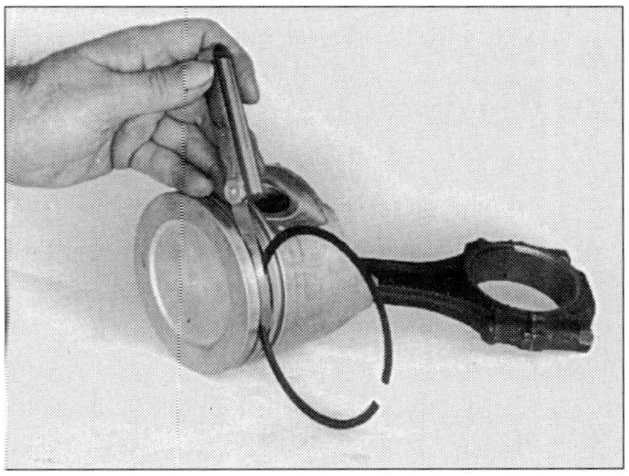

31. Piston ring side clearance should be 0.015 to 0.030-in. for compression rings and 0.002 to 0.005-in. for oil rings with steel rails.

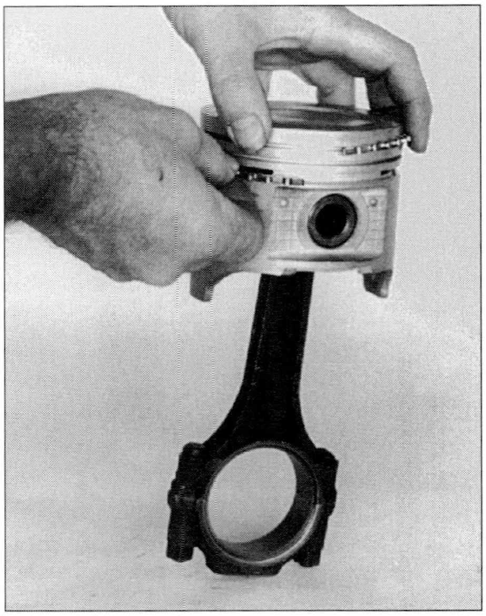

32. Begin ring installation with the oil-ring expander spacer. Be sure the ends butt together and do not overlap.

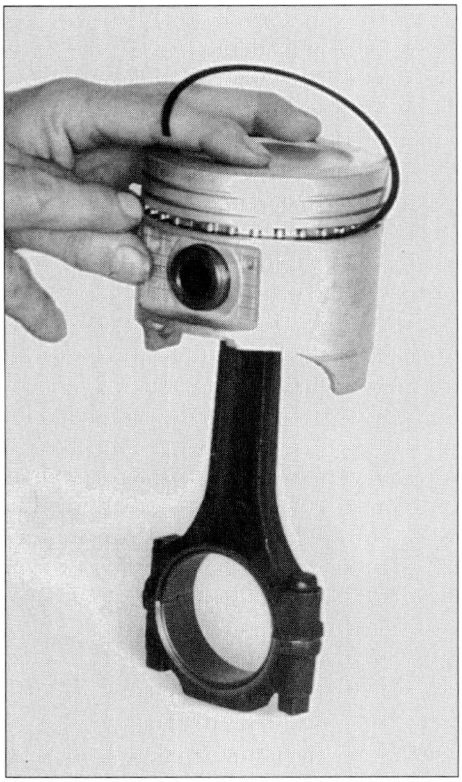

33. Install either the top or bottom side rail. Spread the side rail with your thumbs, put one end in the groove and work the rest of the ring around the piston and into the groove. Be sure the ends of the oil rail are offset from the ends of the oil ring.

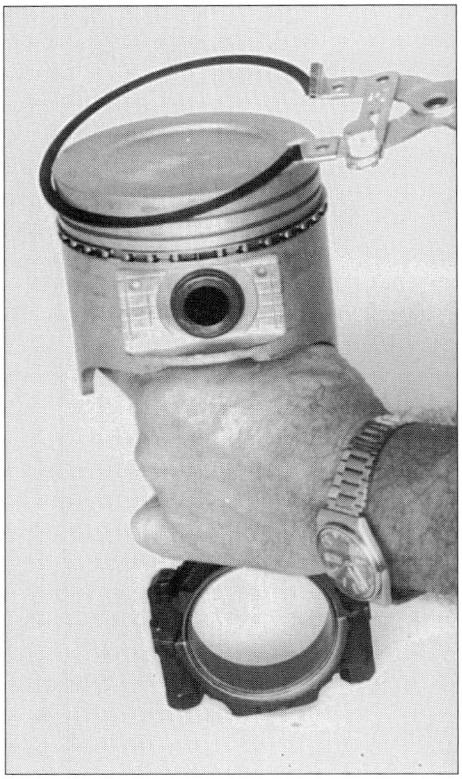

34. Hooray for technology! Use a ring expander to install compression rings #2 and #1. Be sure they're installed per instruction with the ends offset from one another and the oil rings.

35. This is the oiling notch in the connecting rod cap. Remember, it must face inward to squirt oil on its partner.

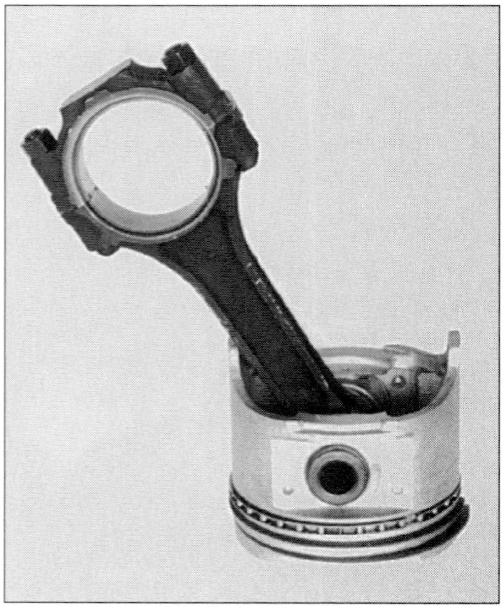

36. The assembled piston ready for installation.

37. Remove only one cap at a time, lubricate the bearings with moly grease and a little oil.

❑ Put a piece of duct tape over the tips of each thumb. Spread the gap with your covered thumbs and bring the ring down into the second groove. Try to do it all in one motion. This helps reduce hanging up in the first groove and eliminates some of the possibility of breaking the ring.

#1 Compression Ring

❑ Check again for up and down.

❑ Expand the ring and install it in the first groove.

❑ Locate the compression ring gaps on opposite sides of the wrist pin. Avoid aligning any two end gaps, oil-to-oil, compression-to-oil or compression-to-compression.

PISTON & CONNECTING ROD INSTALLATION

You'll need the following materials for the next step: a large mallet; a container big enough to hold one quart of motor oil into which you can dip the piston; an oil squirt can; anti-seize compound; rod bolt sleeves; and moly grease.

❑ Assemble tools and lubricants described above. Finding rod bolt sleeves is getting hard to do. I used pieces of 3/8-in. ID rubber hose. You must have something over the ends of the rod-cap bolts to prevent scoring the cylinder walls or crank journals as the rod travels down the piston bore.

❑ Arrange pistons in order of their installation 1-3-5-7 for the left bank and 2-4-6-8 for the right.

❑ Turn the block on the stand so that the right or left bank is in the UP position.

❑ Leave the caps on the piston rods. Remove them only as you install the piston.

38. Fit the rod bolt covers onto the rod bolts. A 3/8-in. ID piece of hose works quite nicely. Fill your container with a full quart of oil and plunge the entire piston into the bath; then let it drain. Wipe the excess around the skirt.

❑ Pour a quart of oil into your large container. I used the bottom half of a one gallon plastic solvent bottle. This will be used to dip the entire top of the piston into for proper lubrication of the rings and wrist pin.

Installation

❑ Rotate the crank until the journal beneath the piston you are going to install is at BDC.

❑ Remove the bearing cap.

❑ Coat both bearing surfaces with moly grease and a few drops of oil.

❑ Install the journal cap protectors.

❑ Check once more for ring gap location. This is the last chance. Adjust accordingly.

❑ Dip your fingers into your "bucket" of oil and coat the inside of the bore, top to bottom.

❑ Plunge the top of the piston into the oil. Let it drain a bit and wipe excess oil around the skirt.

❑ Compress the rings with the ring compressor.

❑ Insert the piston into the bore and level the ring compressor on the block deck. Be sure the arrow or notch faces forward.

❑ With the butt of your hammer handle, begin pushing the piston into the bore. Guide the rod down the bore with your other hand. You may have to use a few raps of the hammer as the piston clears the oil rings and the #2 compression ring.

❑ If anything restricts the movement of the piston, STOP. Find out what it is and correct the matter. It will most likely be a ring hung up on the top of the bore.

❑ When all rings have cleared the ring compressor, finish pushing the piston all the way to the bottom, guiding the rod bolts over the crank journal.

❑ Be sure the rod seats correctly on the rod journal. When seated, give the piston a few raps with the hammer handle to seat the rod bearing.

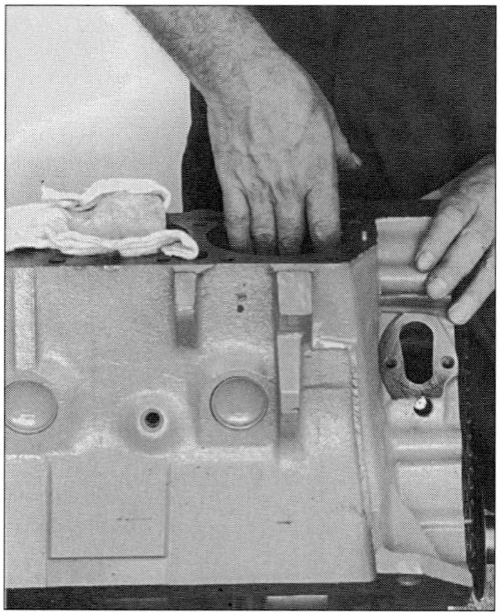

39. To oil the piston bore, grab a "handful" of oil from the container and smear it around the bore. This is finger painting for big kids!

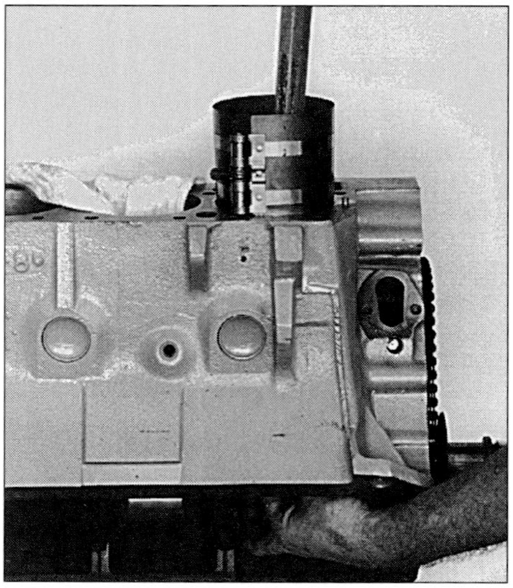

40. I've compressed the rings and am now gently and carefully pushing the piston and rod down through the piston bore. Note how I have my left hand guiding the rod over the crankshaft journal.

41. Down she goes! Notice the notch facing forward. I've also been sure to put piston #2 in the #2 hole. This can be a problem if you start on the passenger side. Remember: 1-3-5-7 driver's side, 2-4-6-8 on the passenger's side.

42. Secure the cap with the oiling hole to the inboard. Just finger-tight for now.

43. While you're down here you might as well install the oil pickup.

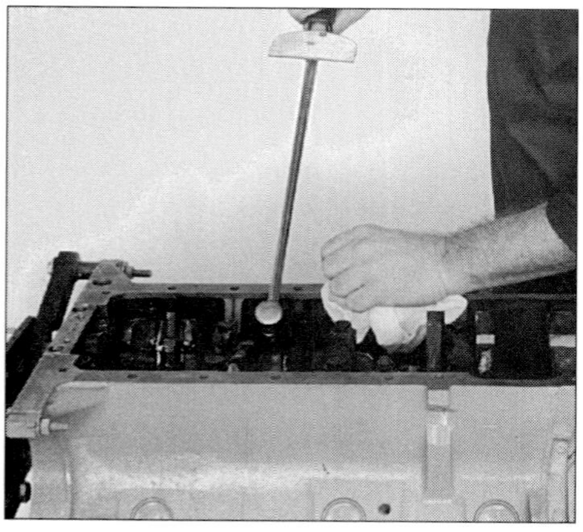

44. When all of the pistons have been installed, torque the connecting rod nuts to 45 ft-lbs. As with the main bearing caps, sneak up on the correct torque a little at a time. Try 25 ft-lbs., then 35 and finally, 45 ft-lbs.

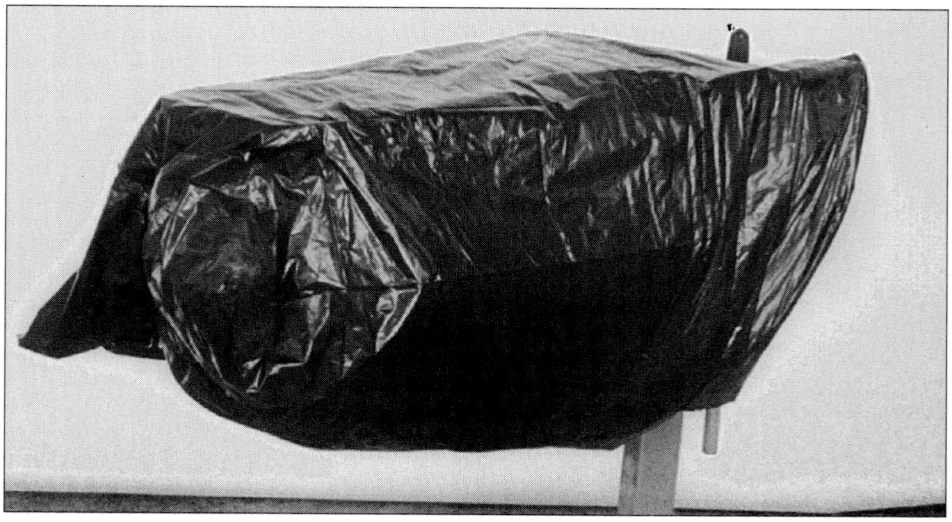

45. I'm tired, it's late and I want to go home—but first I must cover this so someone doesn't inadvertently drop foreign matter in my fine new short-block. A 30-gallon trash bag works best.

❑ Double-check the location of the rod number. It should be pointing outboard. The notch should be pointing inboard. Without this location the engine will not oil correctly.

❑ Turn the block over to access the rod-cap bolts.

❑ Remove the sleeves from the rod bolts.

❑ Install the rod cap with the oil notch pointing to the center or opposite cylinder. 1 should point at 2, 3 should point at 4 and on up the block to 8.

❑ Coat the rod-cap bolts with anti-seize compound.

❑ Install the rod-cap nuts and tighten with a ratchet. Wait until all the pistons and rods have been installed before torquing the nuts.

❑ Torque the rod-cap nuts to 45 ft-lbs. On the 426 Hemi, torque to 75 ft-lbs. Be sure you're using an anti-seize compound here—otherwise you could twist off a rod bolt or tear away the threads through over-tightening.

❑ Check connecting rod side clearance. Clearance, measured with a feeler gauge between two rods, should be 0.009 to 0.017-in. This includes the 426 Hemi.

❑ Make a final test of crank rotation. Everything should turn smoothly and freely. If there are any hang-ups, STOP! Do not force the crank to rotate. Determine the problem and make the necessary adjustments. There should be no problems as you've been rotating the crank all along. However, torquing the rod-cap nuts could change things. Always check for free rotation after each phase of assembly.

❑ Finish this phase of the project by installing the oil pickup. Be sure it's clean and there are no particles caught between the screen and screen cover. Likewise, be sure the pickup tube is clear.

TIMING CHAIN COVER

You're now ready to install the timing chain cover and the harmonic balancer. First, you must install the fuel pump and install a new oil seal in the timing chain cover. The oil seal and the gasket for the fuel pump should be in your gasket kit.

Fuel Pump

❑ Be sure fuel pump is clean and serviceable.

❑ Clean gasket surface by scraping away the old gasket.

❑ Finish gasket surface preparation by sanding with 100-grit sandpaper.

❑ Cement new gasket to pump. Be sure bolt holes are aligned.

❑ Apply a coat of cement to the block-mating surface of the gasket and install the fuel pump to the block. Torque the bolts to 30 ft-lbs.

❑ Rotate the crank to be sure the fuel pump actuator rod contacts the fuel pump actuator arm.

❑ Install the threaded plug that seals the access port to the fuel pump actuator rod. This is located just below the fuel pump body and was the port through which you installed the actuator rod.

Timing Chain Cover Oil Seal

❑ Place the timing chain cover face-down on the bench.

❑ Use a large screwdriver to begin removing the seal from its bore. Insert the blade under the seal and give it a half twist.

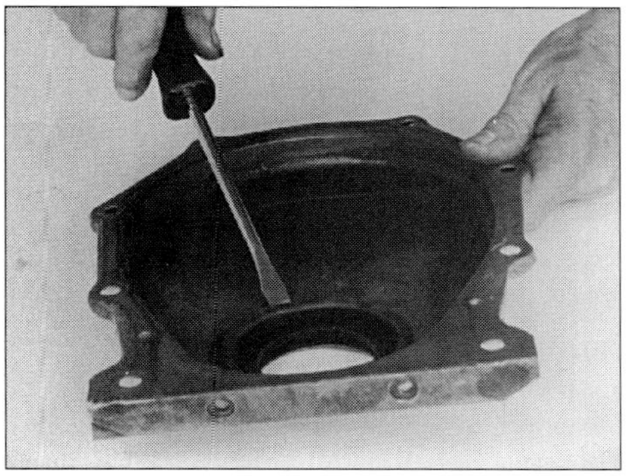

46. Start the front main seal out with a big screwdriver. Usually, a 1/2 turn will do it.

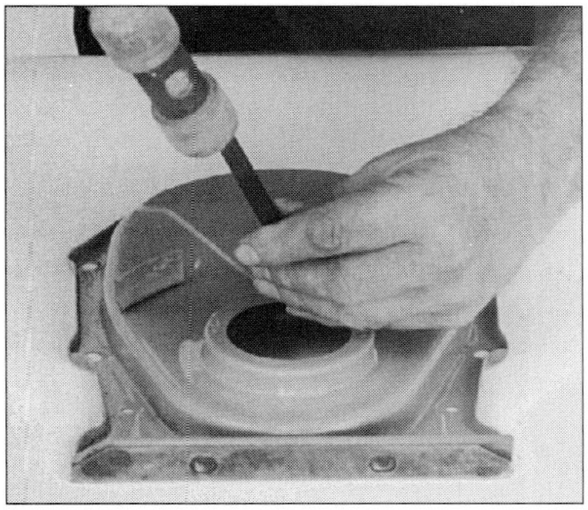

47. Follow up with some punch work from the other side. Work it out little by little all the way around.

❑ Turn the cover over. With a small punch, tap the seal out of its bore starting with the gap created by the screwdriver. Work your way around moving it little by little.

❑ Clean the seal bore.

❑ Apply a thin coat of silicone sealer to the outside of the seal. Be careful not to get any silicone on the neoprene lip of the seal.

❑ Start the seal in with a leather or plastic mallet. Be very, very careful not to bend the edge of the seal. Finish driving the seal into the bore with a large block of wood. Be sure the seal bottoms against the front of its bore.

❑ Put sealer on the lip of the cover and install the gasket. Be sure the bolt holes are clear. Set the cover aside while you prepare the crankshaft damper.

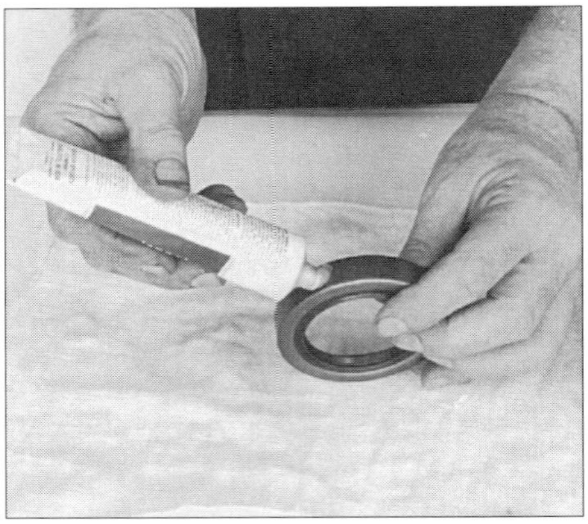

48. Silicone sealer around the outside will insure a leak-free installation. Be very careful not to get any silicone on the lip seal.

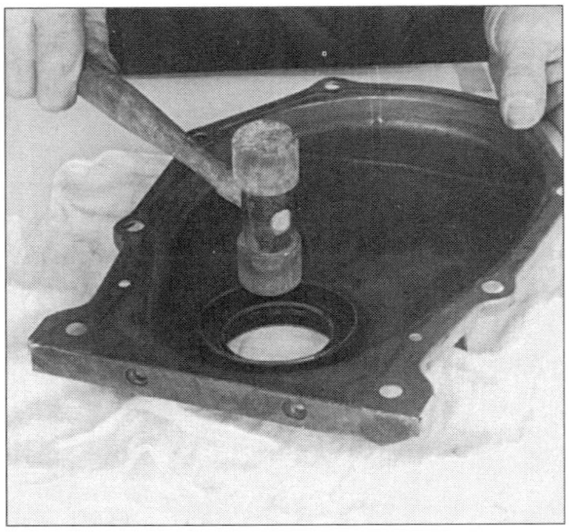

49. Start the seal back in with a mallet. Be sure it is level in the bore and not cocked off to one side or the other.

50. Work your way around tapping it in a little at a time. Be sure it seats well against the bottom of the bore.

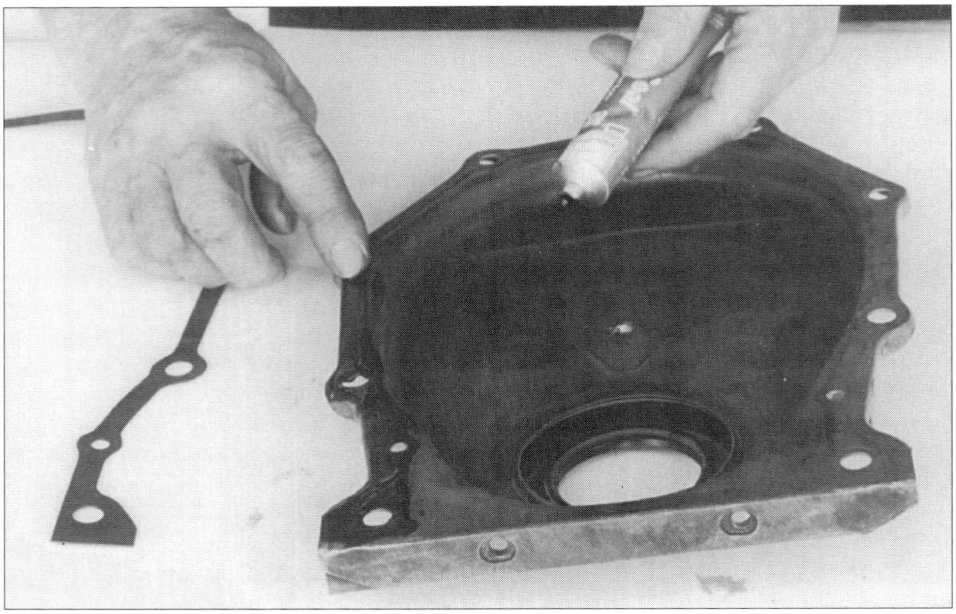

51. Install the gasket with some sealer. Be sure the holes are correctly aligned. Set aside to dry before installation.

52. Install the front cover with the bolts finger-tight. Although there are two alignment pins, there's still a bit of play in the cover. Leave this small amount of play to help center the lip seal around the harmonic balancer.

53. This is very important. Oil both the lip seal and the seal surface of the harmonic balancer. Without this lubrication the front main seal will wear away in the first few revolutions of the engine.

Crankshaft Damper

❑ Inspect the bond between the outer and inner ring of the damper. If there is noticeable movement between the two, replace it.

❑ Clean the inner bore with 400-grit emery cloth. Do not omit this step. If you do, the rust on the bore will cause the damper to seize as it tries to slide along the crank.

❑ With a small file, clean the keyway and remove any burrs.

❑ Clean the damper's outside sealing surface with 400-grit emery cloth. Do not polish. There should be enough "tooth" in the surface to carry oil to the timing chain cover oil seal.

Installation

The secret to a good installation of the timing chain cover and the crankshaft damper is to leave the bolts in the cover loose while you install the damper. This helps center the timing chain cover around the damper insuring long wear.

❑ Be sure the oil slinger is in place. See photo 27.

❑ Select the 8 bolts that fasten the cover to the block. Note the bottom 4 are 3/8-in. and the top 4 are 5/16-in.

❑ Coat the timing cover gasket with sealer and install the cover over the two locating pins on the face of the block.

❑ Install the 8 bolts but only run them in a couple of threads. There will not be much play because of the locating pins but there will be a little.

❑ Be sure the Woodruff key is in place.

❑ Generously oil the lip of the oil seal.

54. After putting a little anti-seize compound on the bore of the harmonic balancer, start it onto the crankshaft with a few blows from the big mallet. Then "pull" it on by threading in the harmonic balancer bolt and tightening.

❑ Coat the inside bore of the crankshaft damper with anti-seize compound. Oil the outside (seal surface).

❑ Slip the damper onto the crankshaft aligning the Woodruff key with the keyway.

❑ With a large mallet (wood, leather or rubber) drive the damper onto the shaft until you can install the bolt and washer.

❑ Insert a block of wood, or as I did, the handle of my hammer, between one of the crank throws and the side of the block.

❑ Tighten the bolt to pull the damper onto the crank and torque to 135 ft-lbs.

❑ Turn the crank over a few times to be sure all is well and aligned.

❑ Tighten the 8 bolts retaining the timing chain cover to 15 ft-lbs.

OIL PAN

Clean the oil pan and get out the oil pan gasket. Mine was made of cork.

55. Torque the harmonic balancer bolt to 135 ft-lbs. Note the hammer handle jammed into the crank to prevent rotation during torquing.

❑ Locate the eighteen 1/4-20 x 3/4-in. bolts that hold the pan to the block.

❑ Check the pan for cracks or holes. Hold it to the light and look for leaks.

❑ Straighten any dimples around the bolt holes created by previous over-tightening. Use your ballpeen hammer. Try to make the surface as flat and straight as possible.

❑ Do whatever is needed to clean the gasket surface. I use a sanding disk chucked into my electric drill.

❑ Apply a medium coat of silicone sealer to the entire gasket surface of the pan.

❑ Set the gasket on the pan and carefully align the holes of the gasket with the bolt holes in the pan.

❑ At the block, trim away the gasket material that extrudes where the timing chain cover joins the block if you omitted this step earlier.

❑ Apply a coat of silicone sealer to the surface of the block.

❑ Set the pan on the block. Finger-tighten the bolts.

❑ Snug the bolts up evenly with your ratchet.

❑ Torque all the pan bolts to 15 ft-lbs.

❑ Be sure now that the drain plug is secure in the pan. Don't dump 5 quarts of oil on the garage floor! Locate the new drain plug seal in your gasket kit, install the bolt and torque it to 20 ft-lbs.

Oil Pump

❑ Fit the oil pump with a new O-ring from your gasket kit.

❑ Put a very light coat of silicone sealer on the mating surface of the oil pump and install the gasket. Use the sealer sparingly. If you use too much sealer it might get ingested into the pump.

❑ Coat the O-ring and the pump shaft with oil.

❑ Seal the block mating surface with a thin coat of sealer and mount the pump. A little twisting back and forth as it goes in will prevent forcing the O-ring out of its groove.

❑ Torque the 4 retaining bolts to 35 ft-lbs.

CYLINDER HEADS

❑ Rotate the block right-side-up on the engine stand.

❑ Remove the heads from their plastic trash bags.

❑ Be sure the dowel pins (head-alignment pins) are in the block deck.

❑ Select a head gasket from your gasket kit. Either one will do as "one size fits all." There is no FRONT or REAR, nor LEFT and RIGHT. Install the gasket on the block deck so that all the holes line up.

❑ Set the head on the block. Be sure the pins and holes align and the head sets flat on the block. Install at least one bolt finger-tight to keep the head from dropping off the block.

❑ Seventeen bolts retain the head to the block; 12 short and 5 long. The five long ones go in the center beneath the rocker arm assembly.

❑ Coat the threads and the areas beneath the bolt head with anti-seize compound.

❑ Snug each bolt down with a ratchet.

❑ Torque all bolts first to 50 ft-lbs. according to the sequence shown on page 134. Then, torque each bolt to 70 ft-lbs. again in the correct sequence.

❑ Thread in a set of old spark plugs to prevent foreign matter from entering the cylinder bores if the engine is going to sit around for any length of time.

Installing 426 Hemi Heads

❑ Repeat the first five steps under "Cylinder Heads."

❑ Be sure the four studs are installed on the valley side of the head.

❑ Install valve lifters (see page 135).

❑ Install pushrods. The short rods in the upper holes (intake) and the long rods in the lower (exhaust) holes.

❑ Head bolts consist of 8 outer cylinder head bolts, 5 rocker arm support bracket bolts and 4 studs with nuts and washers.

❑ Coat all bolt threads and the area under the head with anti-seize compound.

❑ Install the 8 outer cylinder head bolts and snug them with a ratchet.

❑ Install the 4 nuts and washers on the studs protruding into the lifter chamber.

❑ Install the rocker arm support brackets on the locating dowels and engage the pushrods. Install the 5 support bracket attaching bolts. Reread Chapter 6 on *Reconditioning the Heads* to insure the proper location of the rocker arms and rocker arm assembly. This is fairly complex and proper oiling is dependent on everything being in the correct position.

❑ Tighten all bolts with your ratchet.

❑ Torque all bolts and nuts to 50 ft-lbs. in the sequence shown on page 134.

❑ Retorque all bolts and nuts to 70 ft-lbs. following the same sequence.

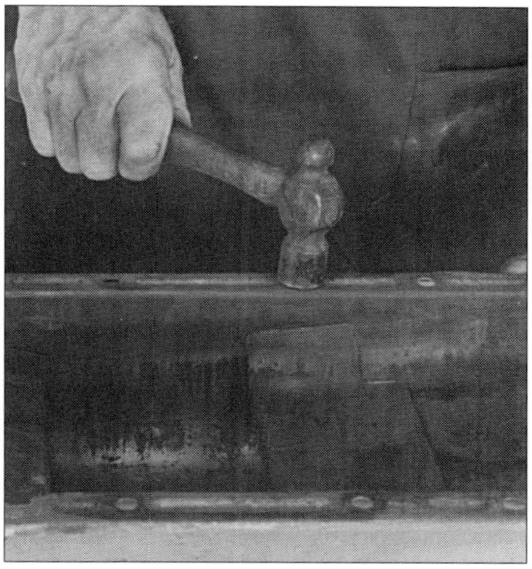

56. If there are any dimples around the oil pan bolt holes from excessive tightening, flatten them out with a hammer. You may need to use something to support the lip of the pan from beneath. I find a 2x4 to be quite satisfactory.

57. This arrangement works well to clean up the gasket surface. If you don't have a disk sander, go back to the sanding block. Just get this sealing area really clean.

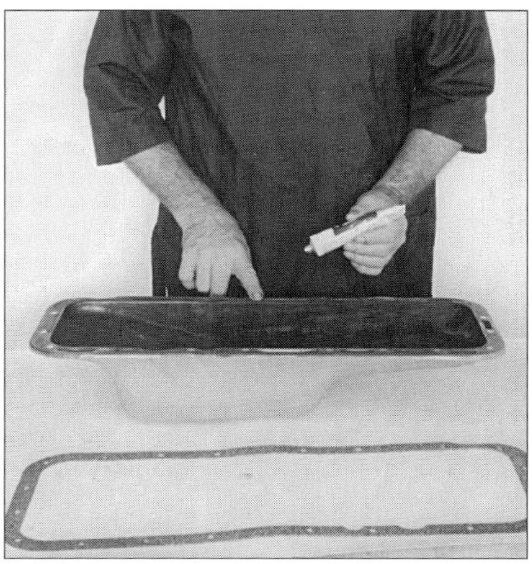

58. Good old silicone sealer: where would we be without it!

59. Here's an easy problem to forget. The front cover gasket protrudes above the machined surface just a bit. Level it off with a single-edge razor blade.

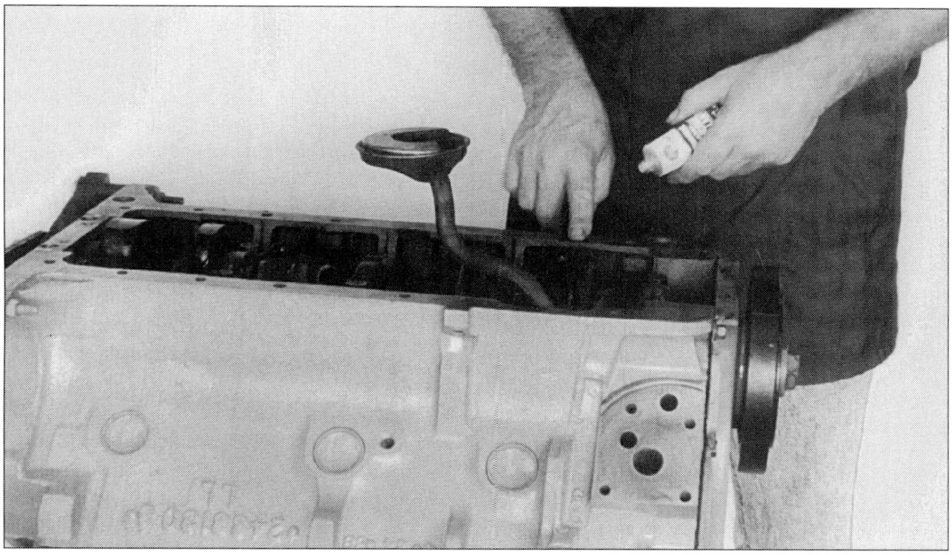

60. Seal the surface of the block . . .

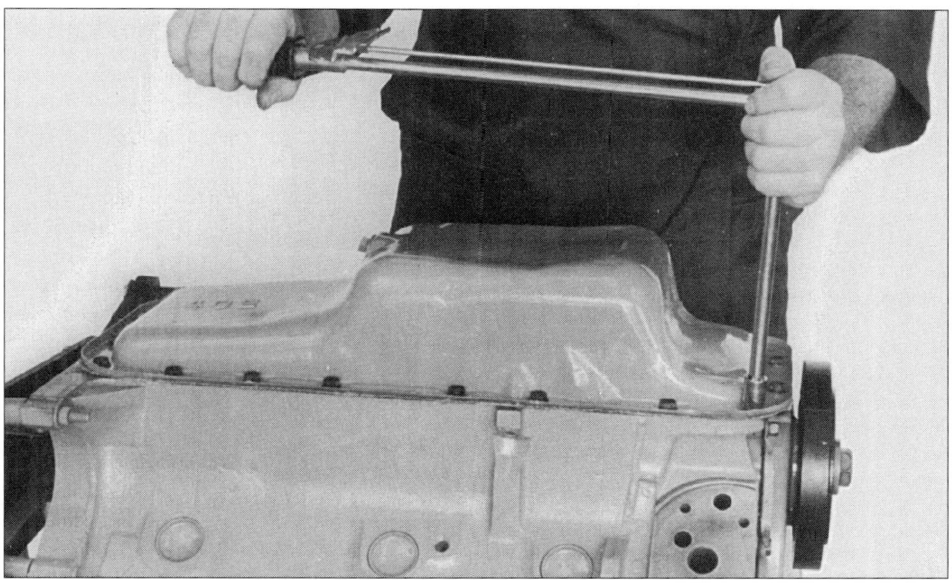

61. . . . and install the pan. Tighten all of the bolts, alternating sides to even the pressure, to just a bit more than finger-tight. Then torque all of the bolt to 15 ft-lbs. using the same alternating pattern.

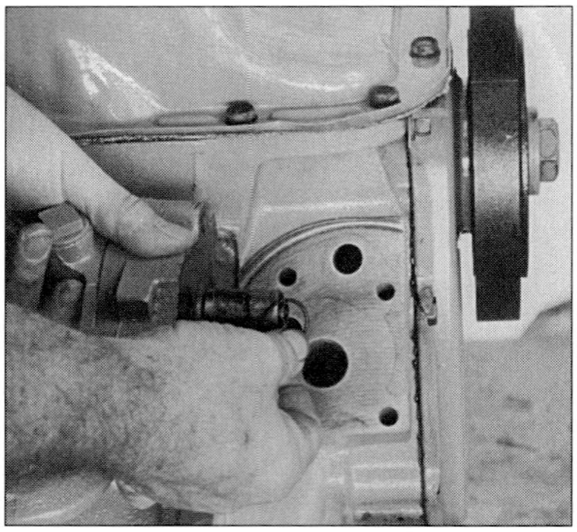

62. Your gasket kit will include all the necessary O-rings. Install a new one on the oil pump.

63. Lubricate the O-ring to keep it from possibly twisting out during installation.

64. Four bolts secure the oil pump to the block. Torque to 30 ft-lbs.

65. After setting a head on the block, thread in at least one bolt to hold it in place. I use two here. Once I had a head rotate around the one bolt supposedly holding it in place. The side loading cracked the head at the bolt hole. Now I generally use two bolts.

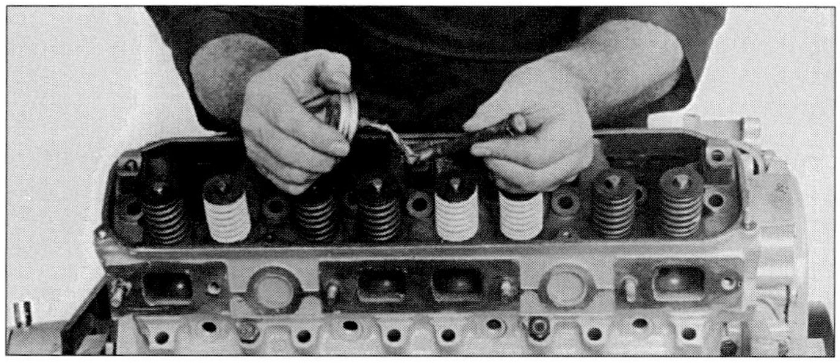

66. Lubricate the threads and the shoulder area beneath the head of the bolt before installing—again to assist in accurate torquing.

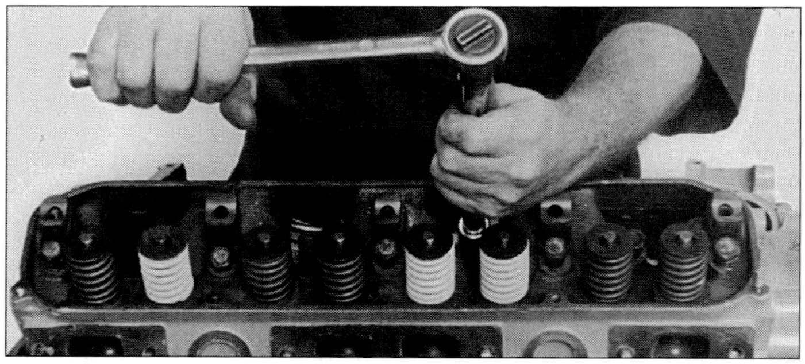

67. Snug up all the bolts with your big ratchet. I'm using a 1/2-in. drive here. Then, sneak up on that torque of 70 ft-lbs. Be sure to follow the bolt-tightening sequence shown in the drawing below left.

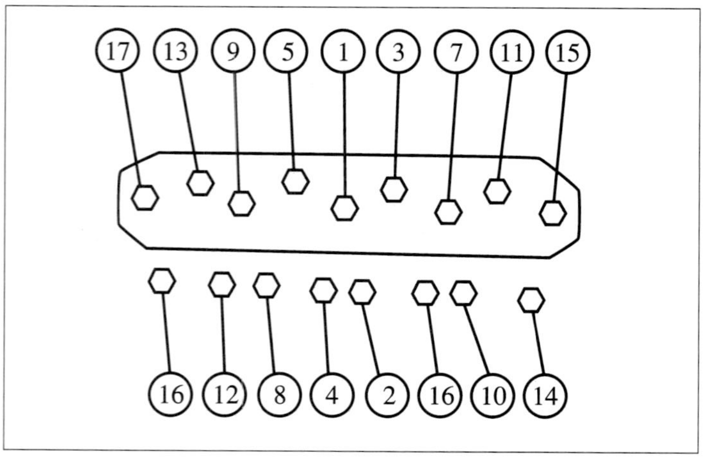

This is the cylinder head bolt-tightening sequence for all big-block engines but the 426 Hemi. Follow this sequence religiously; proper sealing and head-life depend on it.

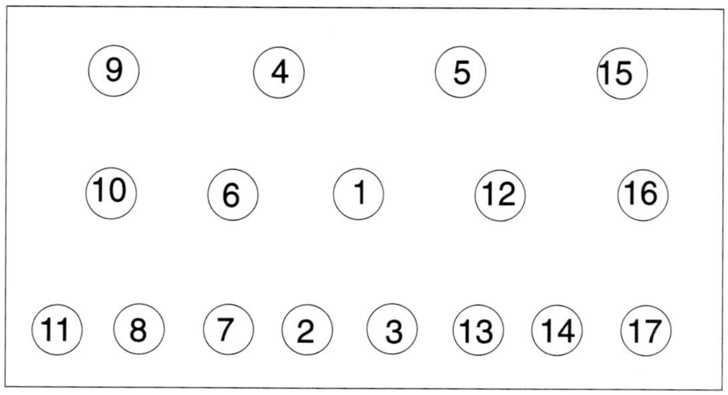

To accurately torque the head bolts on your 426 Hemi, use this tightening sequence. Numbers 9, 4, 5 and 15 are the studs that must be accessed from inside the valley.

FINDING TOP DEAD CENTER (TDC)

In the four-cycle engine there is one TDC but it occurs twice in each full cycle: once at the end of the compression stroke and once at the end of the exhaust stroke. Right away this suggests something to you. On the compression stroke both valves will be closed; on the exhaust stroke the exhaust valve will be open as the piston moves up.

Use a piece of mechanic's wire stuck through the spark plug hole to feel the top of the piston. Rotate the crank with your breaker bar and feel the piston rise in the bore. The piston will be at TDC on the compression stroke when you feel it there with your wire and both valves are fully seated. (The rocker arms will be unloaded and you should be able to jiggle them a bit.) The final check is to note the location of the timing mark on the harmonic balancer. It should be pointing to the "0" on the timing scale.

I've inserted a piece of coat hanger wire through the number 1 spark plug hole. With it, I can "feel" the top of the piston as it moves up and down.

If I'm checking timing I want the harmonic balancer to "zero" when piston number 1 is at TDC on the compression stroke. If it "zeros" at TDC on the exhaust stroke I'm 180° off.

❑ Thread in a set of old spark plugs to keep foreign matter out of the engine if it's going to sit for a length of time.

VALVETRAIN

❑ Rocker arm assemblies should be together and ready to install. Refer to page 106 if you have not already done so.

Valve Lifters

❑ If you're using the old valves they must be returned to their original locations.

❑ New valve lifters may be installed in any valve lifter bore.

❑ Prime hydraulic lifters with your oil squirt can by forcing oil through the hole in the side of the lifter until it comes out the top.

❑ Place a few drops of oil in the top of the lifter and thoroughly oil the body. Dab some moly grease on the foot.

❑ Oil the entire bore of the lifter by squirting oil on your finger and rubbing it around the circumference of the bore.

❑ Insert the lifter into the bore. Add another squirt of oil between the bore and body of the lifter.

❑ Slide the lifter into the bore. Spin it around a few times to insure complete oiling.

Pushrods

❑ Check the pushrods for straightness by rolling each one over a known flat area. A piece of window glass works well. If at any time you can insert a 0.020-in. feeler gauge between the center of the pushrod and the surface, replace the pushrod.

❑ Insert the pushrod through the head and into its lifter. New pushrods can go in any direction. Used pushrods should be reversed from their original position. Check the wear pattern at each end. The worn area on the lifter end will be much smaller than the rocker end.

Rocker Arm Assemblies

❑ Set the rocker arm assembly onto the mounting stand. You marked each side right and left. If you failed to do so refer to Chapter 6 on *Reconditioning the Heads* to be sure the rocker arms are in the right position, the oiling holes are pointed in the right direction and everything is in order.

❑ On all but the 426 Hemi there are 3 narrow spacers and 2 wide spacers accompanying the 5 rocker arm retainer bolts. Two of these bolts are longer than the other 3 and accompany the wide spacer. The 3 narrow spacers go at each end and in the center. The 2 wide spacers go over the area between.

❑ Set the rocker arm assembly on its bosses. Thread in the five bolts and spacers by hand.

❑ Adjust the pushrods to fit in the end of the rocker arms.

❑ Begin tightening the rocker arm bolts a little at a time, pulling the assembly down to seat on its bosses.

❑ Be sure the pushrods remain in place—both under the rocker arm and in the valve lifter.

❑ Torque the bolts to 25 ft-lbs.—30 ft-lbs. for the 426 Hemi.

❑ Thoroughly oil all the moving surfaces of the assembly.

❑ The following is of the utmost importance. With your breaker bar and 1-1/4-in. socket, rotate the crankshaft through a couple of 360° rotations to be sure everything is moving freely. It's easy to have a pushrod jump out of position while tightening the rocker arm assembly. If you catch it now it's simple to fix. If you wait until the engine is in the car and you try to start it, you'll bend the pushrod. Check now rather than pay later.

Initial Valve Adjustment (Solid Lifters)

❑ Rotate the crankshaft until piston number-1 is at TDC at the end of its compression stroke—both valves closed and the timing notch on the damper is pointing to the O on the timing marks on the timing chain cover.

❑ Make a chalk mark 180° from the notch on the damper (directly opposite on the other side).

❑ Adjust the lash on each valve according to the valve lash chart below.

OIL PUMP DRIVE

❑ Bring number 1 piston to TDC.

❑ Insert the oil pump drive into the distributor hole, making sure the hex shaft engages the oil pump and the gear drive meshes with the cam gear.

❑ Slot in the center of the gear must align with the centerline of the camshaft. This requires inserting the drive one or two teeth off so that it will rotate into correct alignment as it slides into the bore.

INTAKE MANIFOLD (383, 400, 413, 440)

❑ Select either new valley cover gasket style or, if you plan to use your old gasket, new sealing gaskets.

❑ Coat all mating surfaces of the heads with a thin coat of gasket sealer that comes with the gasket kit. Otherwise use one of the Permatex type sealers. Seal the front and rear edges of the block with silicone sealer.

❑ Coat the mating surfaces of the intake manifold with the same sealer you used on the heads.

❑ Set the manifold in position and thread in the manifold bolts. Do not tighten.

❑ Mount the front and rear cover bars and install the three retaining bolts.

❑ When you're sure all is seated well, remove the left front and rear bolts and mount the front throttle spring bracket and the throttle bellcrank in the rear.

❑ Torque manifold bolts to 50 ft-lbs.

❑ Torque cover bar bolts to 15 ft-lbs.

Intake Manifold (426 Hemi)

❑ The intake manifold is attached with two studs, four 1/4-20 x 2-1/2-in. screws, six 1/4-20 x 2-3/4-in. screws and four 1/4-20 x 3-in. screws.

❑ Install a stud in the right front and left rear locations. Use the two air cleaner studs in the left front and right rear

VALVE LASH

	INTAKE	EXHAUST
	0.028	0.032
❑ With number-1 at TDC, adjust:	2 and 7	4 and 8
❑ Rotate crankshaft 180o clockwise with #4 at TDC, adjust:	1 and 8	3 and 6
❑ Rotate crankshaft another 180o with #6 at TDC, adjust:	3 and 4	5 and 7
❑ Final rotation of crankshaft 180o with #7 at TDC, adjust:	5 and 6	1 and 2

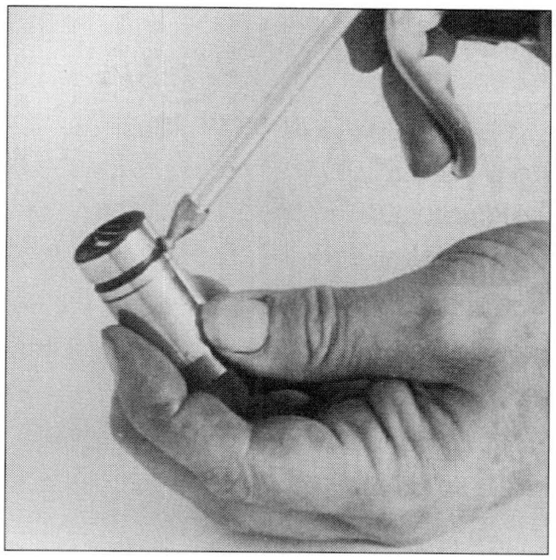

68. With some judicious squirting you can get enough oil into the lifter to prime it prior to starting the engine. There is a hole inside this oiling groove.

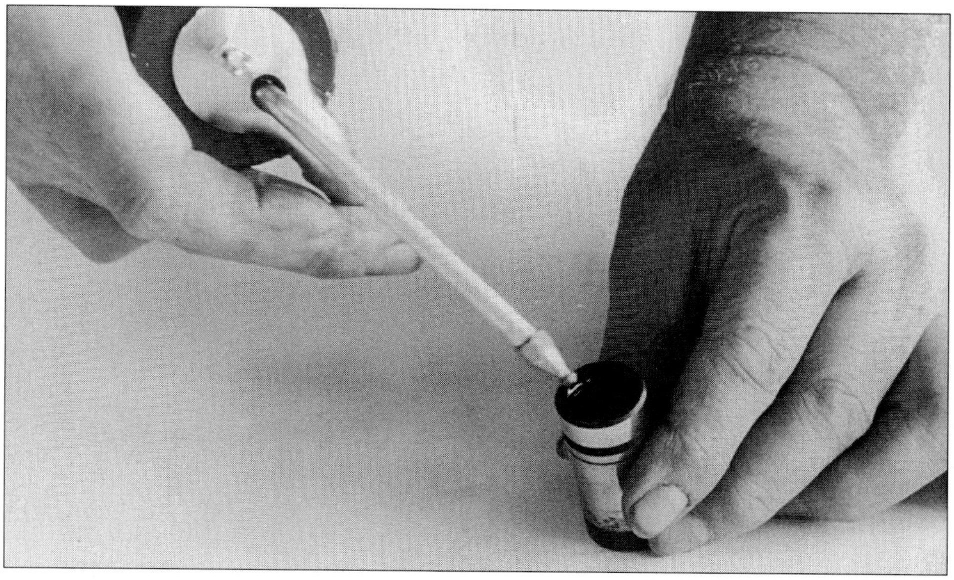

69. Fill this cup with oil to lubricate the bottom of the pushrod.

70. Lubricate the lifter bores with plenty of oil. Spread it with your finger.

71. Hold the lifter in place and squirt in a little oil as you drop the lifter into its bore.

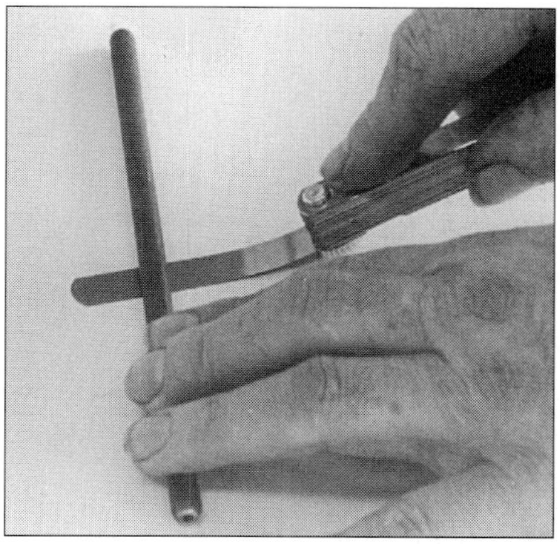

72. The pushrods must be perfectly straight. Replace if there is more than 0.020-in. bend in it.

73. Drop in the pushrods before you try to mount the rocker arm assembly. Be sure each is seated in its respective lifter.

locations to hold the gaskets in place when setting in the manifold.

❑ Run a bead of silicone sealer around the water ports of the heads.

❑ A bead is stamped around all the ports of each metal gasket. These are set in place with the bead against the head.

❑ Apply a coat of Fel-Pro Pro-loc or silicone sealer to the valley-pan-to-block surfaces and to the mating surface of the block. Set the pan over the locating dowels (three at each end).

❑ Apply silicone sealer around the water ports on the intake manifold.

❑ Set the intake manifold in place. Thread in the screws (with threads coated with anti-seize compound) in the following manner: the four shortest screws (2-1/2-in.) in the four forward holes of the manifold. The four longest (3-in.) go in the four rearmost holes. The six remaining 2-3/4-in. go through the carburetor ports.

❑ Refer to the diagram above for the tightening sequence. Torque the bolts marked "B" first to 6 ft-lbs. then the bolts marked "A" to 4 ft-lbs. The figures of 4 and 6 ft-lbs. are not misprints. The heads can be damaged by overtorquing these screws.

VALVE COVERS

❑ Clean the valve cover gasket surfaces.

❑ Apply a medium coat of silicone sealer to the gasket surface.

❑ Install valve cover gaskets. Be sure the locating tabs are in place.

❑ Set the valve covers aside to let the silicone dry.

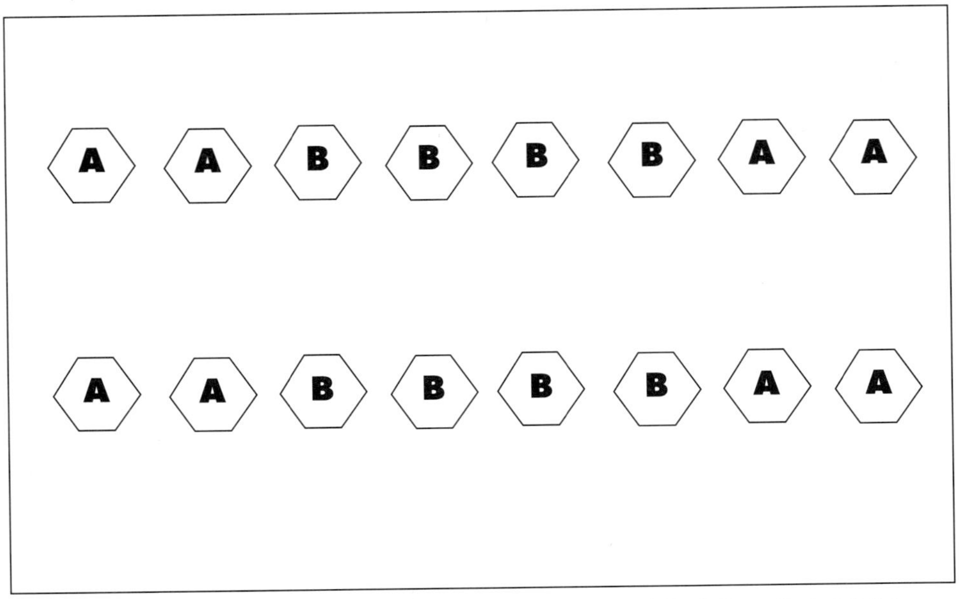

Torque the bolts of the intake manifold in the sequence above. Bolts "B" are first at 6 ft-lbs., bolts "A" are second at 4 ft-lbs.

❑ When the silicone sealer is dry, mount the valve covers to the engine without further sealer. Engines with adjustable rocker arms will have to have the valve covers removed to access the rocker assemblies. After the engine is in the car you may need to remove a valve cover to check on operations—so don't seal the valve covers to the heads.

❑ The valve cover with the oil filler tube goes to the driver's side.

❑ Torque the passenger side bolts (and nuts) to 5 to 7 ft-lbs.

❑ Leave the driver's side until after installation of the distributor. You'll need to remove the cover to help locate TDC of the number 1 piston.

EXTERIOR HARDWARE

The exterior hardware includes the throttle bellcrank, the front throttle spring bracket, spark plug wire harnesses and spark plug heat shields. The efficient mechanic will have installed these as he or she installed the heads, intake manifold and exhaust manifolds. However, this gets a bit confusing to the first time engine builder. Therefore, I am including all of the above as exterior hardware and will describe its installation "after the fact." Additionally, to make your life even easier, I have painted these parts black to stand out in contrast to the engine. If you're doing a restoration these parts should be painted the same color as the block.

Water Pump Housing

❑ Clean gasket surfaces.

❑ Install gasket from kit with silicone sealer.

❑ Coat block-to-housing surface with sealer.

❑ Mount housing with two bolts: one behind the front heater-hose pipe and one on the opposite side between the A/C mounting bosses. The remainder of the mounting bolts will be added after the engine is in the car.

❑ Torque the two bolts to 30 ft-lbs.

❑ The water pump itself is left until the engine is in the car.

Thermostat

❑ Install the thermostat gasket on the water pump housing with silicone sealer.

74. Install the rocker arm assembly. Tighten a little at a time to prevent serious loading of the rocker shaft. Keep your eye on the pushrods—they tend to jump out of the lifters sometimes.

75. Oil all the contact surfaces before installing the valve covers.

76. The intake manifold gasket/valley cover comes with its own sealer. Use it liberally to seal around the head ports.

77. Use your own silicone sealer to get this area. Place a generous glob at the mating edge between the head and block.

78. Carefully set the gasket/cover in place.

79. Set the manifold in place then secure the front and back of the cover with the cover straps. Install the intake manifold bolts and torque to the specifications listed nearby for the 383 through 440 and for the 426 Hemi.

80. The water pump housing is retained with two bolts at this time: one just behind the front heater hose pipe and one on the driver's side between the power-steering-pump mounting bosses. The remaining mounting holes are bosses for further accessories.

81. Always install a new thermostat. The gasket goes down first and is sealed to the water pump housing. Set the thermostat on top of the gasket, apply some silicone sealer to the thermostat housing and mount—torquing the two bolts to 15 ft-lbs.

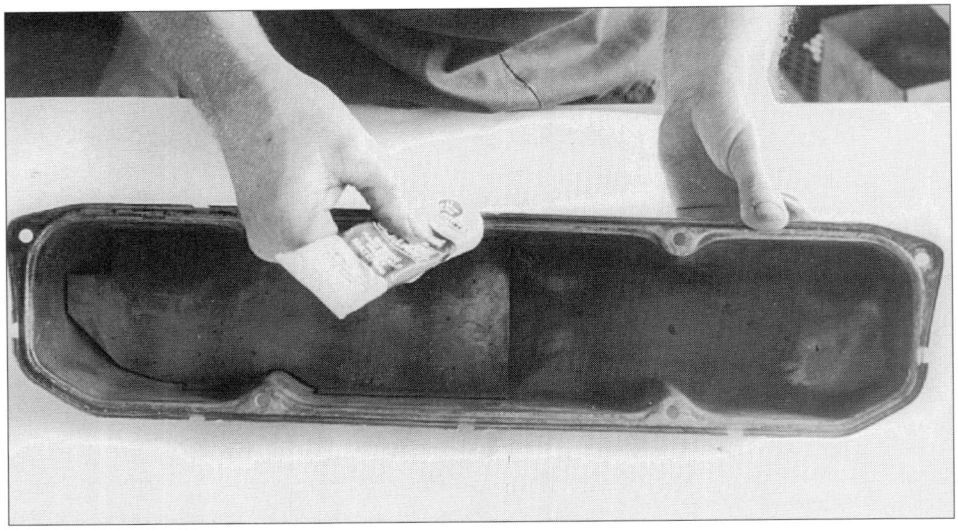

82. Clean and seal the gasket surface of the cylinder head cover (valve cover).

83. Install the gasket. Be sure the tabs you see here are in place, correctly holding the gasket while the sealer drys.

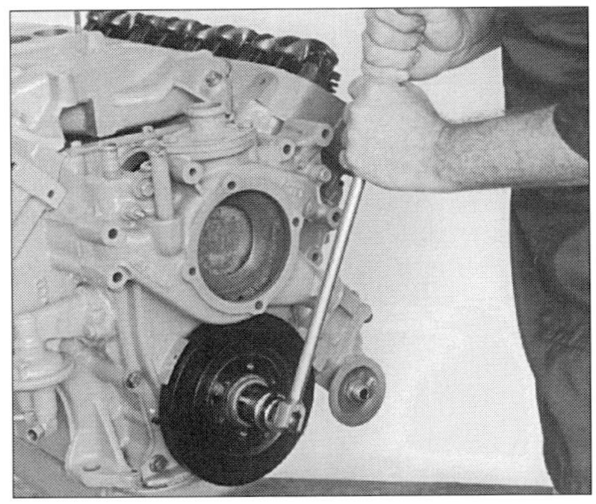

84. While you're waiting for the sealer to dry on the cylinder head cover gaskets, try rotating the crankshaft one more time to be sure nothing is hanging up. It sure is a disappointment to get the engine back in the car and find you left your shoe in the number four cylinder.

85. Be sure this spacer is in place between the water pump idler pulley and the water pump housing. If you omit this part the idler won't align with the crankshaft pulley.

86. Here, we are all set to tighten the water pump fanbelt.

87. This photo will help you get the alternator mounting bracket in the right position.

❑ Set thermostat on top of the gasket with the pointed end up.

❑ Apply silicone sealer to the mating surface of the thermostat housing and install.

❑ Torque the two mounting bolts to 15 ft-lbs.

Water Pump Idler Arm

❑ Mount the water pump idler arm in the two holes to the right of the water pump mounting flange.

❑ Don't forget the 2-in. spacer. The square 1/2-in. hole at the top allows you to use a 1/2-in. breaker bar to adjust belt tension.

Alternator Bracket

❑ Mount the alternator bracket to the right side head as seen in the photo. This bracket can be mounted several ways. The correct way is for the "dog leg" to be in the "up" position.

❑ Torque bolts to 30 ft-lbs.

Engine Mounts

I hope you remembered to mark the engine mounts "right" and "left" or "driver/passenger." If you didn't, I'll try to describe the difference. In the area of the mounting holes, the "arm" with the single mounting hole is longer for the passenger side. On the side with the two mounting holes the cutout at the top leaves a "lip" at the top for the driver's side. Photos 90-93 should help you get things right. Attach the mounts to the rear faces of the mounting bosses with the bolts passing through from the inside toward the outside for the two top bolts.

❑ Torque to 45 ft-lbs.

Dipstick Tube

❑ Mount the dipstick tube at the same time you install the engine mount bolts at the driver's side. The mounting bracket for the dipstick tube attaches to the top rearward bolt of the engine mount.

❑ Spread a little silicone sealer around the base of the dipstick tube where it enters the block.

Throttle Bellcrank and Throttle-Spring Bracket

❑ The throttle bellcrank mounts to the rearmost bolt hole of the intake manifold on the driver's side.

❑ The throttle-spring bracket mounts to the forward most hole of the driver's side intake manifold. If your car is equipped with an air conditioner, the throttle-spring bracket will mount to the top of the air conditioner support leg which also uses this mounting hole.

❑ Torque to 50 ft-lbs.

Spark Plug Wiring Harnesses and Heat Shield

❑ The spark plug heat shield mounts between plugs number 6 and 8 (passenger side) under the two head bolts.

❑ A two-wire harness mounts under the head bolt to the rear of number 2 spark plug.

❑ There are two, four wire harnesses. The long thin one mounts to the first exhaust manifold bolt on the passenger side. Mount with the "long" part in the "up" position.

❑ The four-wire harness with the large, flat mounting surface attaches beneath the driver's-side bolt on the front valley cover sealing strap.

❑ The final, two-wire harness mounts to the number 3 head bolt on the driver's side just in front of the number 3 spark plug.

❑ Retorque head bolts to 70 ft-lbs.

Exhaust Manifolds

❑ Be sure all studs are in place and the threads are clean.

❑ Install the two flange gaskets from your gasket kit. These are self-sealing and require no other sealers.

❑ On the driver's side, the next to last mounting hole on the exhaust manifold requires a 3-1/2-in. shoulder bolt. This is seen in photo 95.

❑ On the passenger's side, don't forget the wiring harness that mounts to the front bolt.

❑ Mount the exhaust manifolds and torque the nuts to 30 ft-lbs.

SUMMARY

If some serious luck has been with us, we now have the engine back together and ready to shoehorn into the car. I hope the format change was helpful to you. Likewise, I hope the demonstration attaching some of the exterior hardware after the fact, was more helpful than confusing.

Check the engine over one more time. Look especially carefully at all the core plug areas—water and oil. As you saw in the photos, I forgot a couple in the natural sequence of things. I caught them in time by taking my own good advice.

Keep a shop towel shoved into the carburetor inlet and in the distributor mounting hole. This will prevent small parts from accidentally falling into the interior of your new engine.

Congratulations on making it this far! Now, let's get this rascal into the car. ■

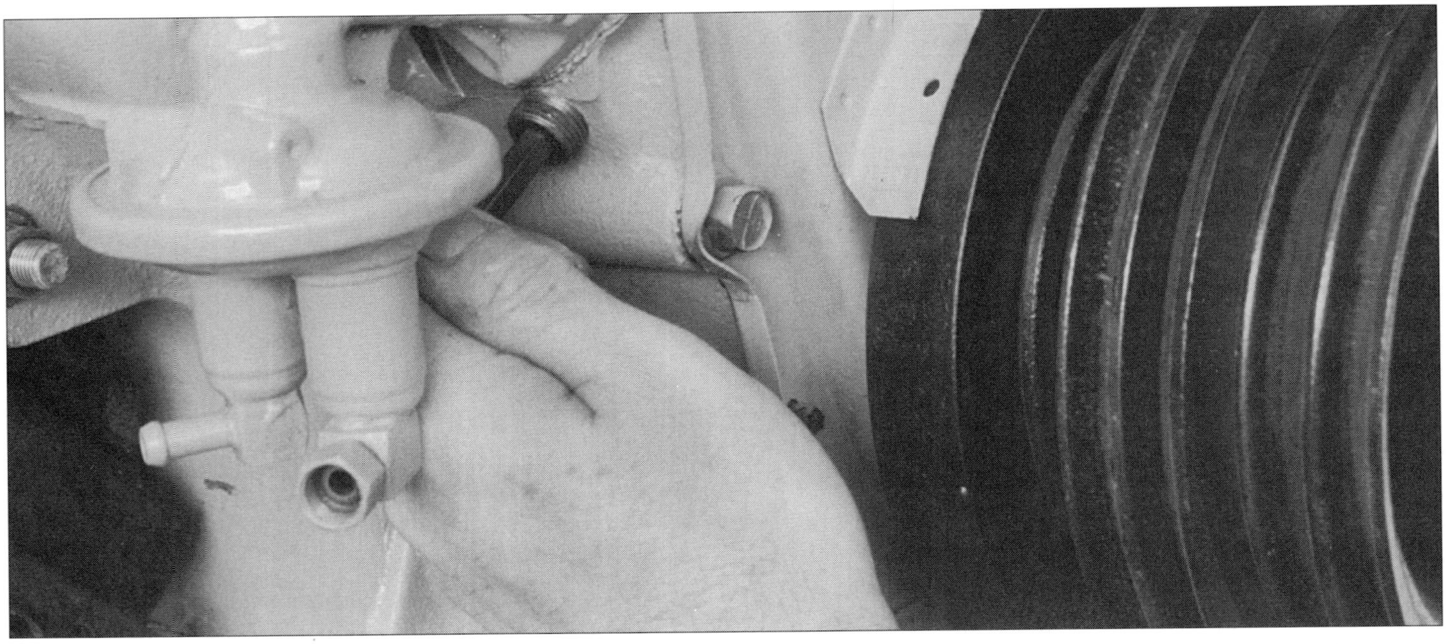

88. Everyone makes a mistake! I left this access hole to the fuel pump actuator rod unplugged until almost too late. Had I forgotten it entirely, oil would have blown out of here like Old Faithful.

89. Some more help in mounting the accessories. The throttle bellcrank goes under the rearmost bolt of the intake manifold.

90. For a clear picture I have assembled the driver's side engine mount without the oil dipstick tube. Photo 93 shows the correct mounting with the dipstick in place.

91. Here's the passenger side engine mount. Can you see the long "ear" of the rearmost mounting flange? This helps define the passenger side from the driver's side.

92. This is a nice clean "shot" of the spark plug heat shield and the spark plug loom mounting points.

93. If your engine mount, dipstick tube and spark plug wire loom look like this on the driver's side, chances are you have everything together correctly.

94. These are the mounting points for the throttle spring bracket and driver's side wiring harness.

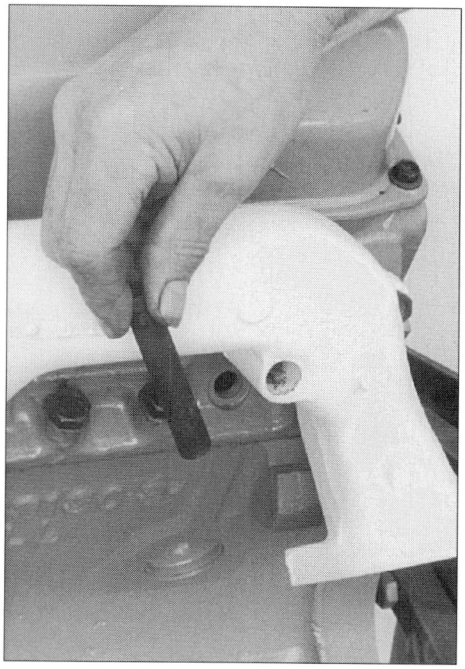

95. This shoulder bolt is a "must" for mounting the driver's side exhaust manifold. If you've lost it, you'll have to order another from Mopar Performance.

96. A nice view of the completed engine. Other than the strange looking black parts, I think it's great. Usually, painting is left until everything is assembled but the engine mounts and exhaust manifolds. Then, the entire block is painted, usually with the correct factory paints for authenticity.

INSTALLATION

8

Well, it's time to put it back in. All you have to do now is find all the parts. Where's that radiator shroud? For that matter, where's the radiator? Oh, yes, it was sent out for a new core. The shroud is probably behind the water heater.

Hopefully, you followed all the directions I gave about engine removal, so everything is stored away. It's marked so you'll know where it goes. You even know where you placed all the Polaroid shots you made of things you might forget! Now you're really going to be happy you took that extra time to carefully organize the project.

Many of the parts are still greasy so let's go to work on them. Solvent, soap and water and lots of scrubbing should get everything clean enough. A couple of aerosol cans of black or gray paint will go a long way toward making everything look sharp again. Of course, if you're working toward a factory original restoration, each of those parts must be disassembled, bead blasted, painted or plated in accordance with the original.

Most of us will be happy just scrubbing out the engine compartment to get the grease we missed when we steam-cleaned the engine before removing it. If you want to go a bit further, mask off all the lines, leads, hoses, and hardware then spray the engine compartment the same color as

Because you don't have to pay family members to model, I've called on my brother Al to pose for our pictures—and to help. Here he's started to remove the torque converter to access the transmission front seal.

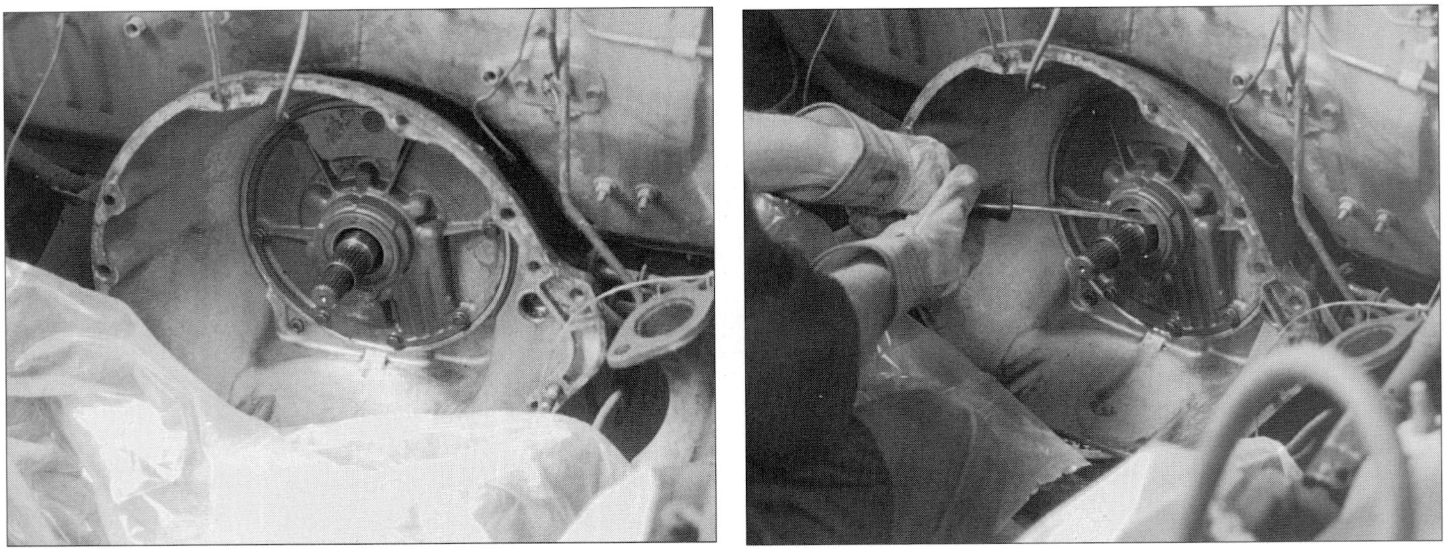

With the torque converter out, it's easy to get to the front seal. Al is going to pry it out with a screwdriver. The billowy white material you see in the photo above is visquene—a plastic film Al put down so he doesn't get the seat of his pants greasy.

the body. This is the way it was done at the factory.

Now, collect your tools, get a hoist, find the jack stands and jack and let's put this beauty back into its home.

AUTOMATIC TRANSMISSION

The first thing you must do is replace the front pump seal in the transmission. This is a must. Murphy's Law knows it will start to leak as soon as you put the engine back in.

The front pump seal looks just like the timing chain front cover seal—they're both lip seals. To reach the seal, remove the torque converter—the big bowl-shaped gizmo inside the bellhousing.

The good-looking fellow you see in the photograph removing the torque converter is my brother, Al. Al can make anything run; an automobile engine or a juke box—if he can tweak it, it'll run better. In his early years he was an up-and-coming racer. Today, he builds and restores a wide variety of interesting automobiles and race cars. Right now, he's my "wrench" while I do the photography.

Slide the torque converter off the input shaft—it will take a little wiggling. Be careful! It's full of transmission fluid. Tilt it forward to keep from spilling ATF all over the place. Pour the ATF out once the converter is clear of the car.

With the converter off, put a screwdriver behind the seal lip and pry it out. Pry a little at a time around the seal until it pops out. Don't use a hammer and chisel, because you could easily damage the stator support or the seal bore.

Run a bead of silicone sealer around the edge of the new seal. Wipe the seal bore clean with lacquer thinner. Oil the lip seal as you did for the front cover. Place the seal over the input shaft and square it in the bore with the seal lip pointing toward the transmission. Use the old seal as an installation tool and gently tap around the circumference until

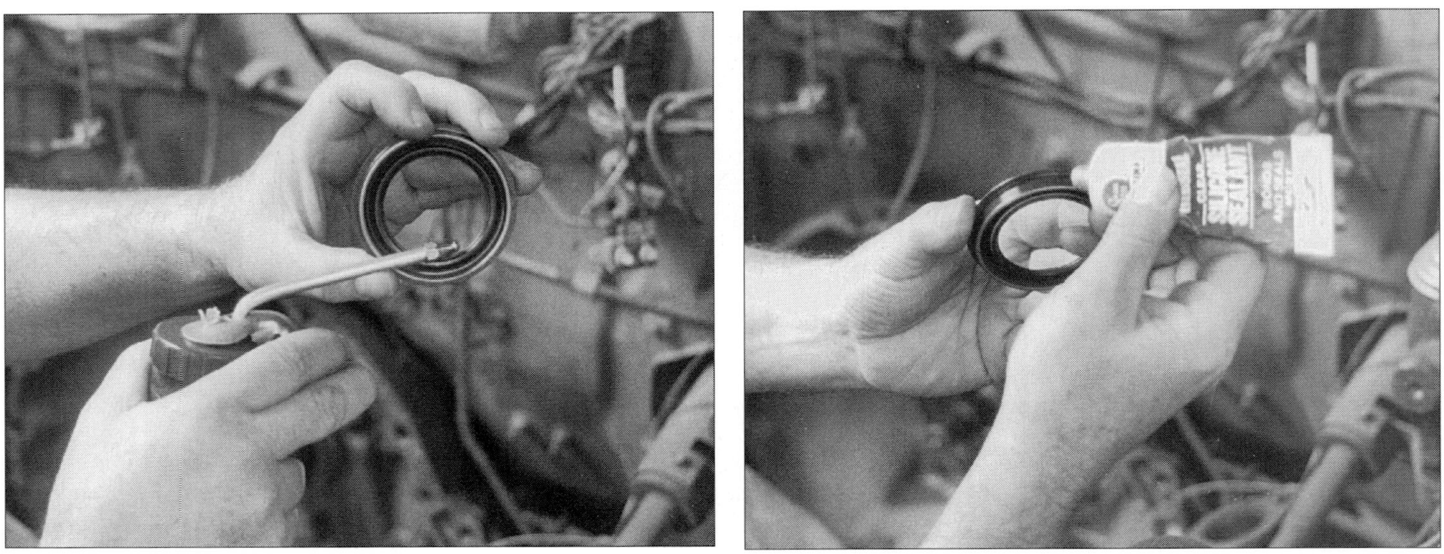

As you have done with all the seals and gaskets, seal the mating surface and lubricate the contact surface of the transmission front seal.

Al uses a 2x4 to spread the force as he taps the seal into its bore. He could have also used the old seal to hammer on. This, too, would prevent denting the new seal.

it bottoms in the bore. Al uses a piece of 2x4 to spread the force.

Wipe off any excess sealer. Clean up the inside of the bellhousing and inspect the bellhousing-to-block mating surface for nicks or burrs. Check the outer surface of the spline for rust, burrs, scratches or nicks. These can be removed with 400-grit emery cloth.

Coat the input shaft and converter "snout" with oil, then slip the converter onto the shaft. The converter must engage the transmission-input shaft, the front pump and stator support—three distinct engagements.

Rotate the converter while pushing on its nose until you think it's fully engaged. Lay a straight-edge across the face of the bellhousing and measure from it to one of the lugs on the converter. The distance should be about 1/2-in. If it's less, the converter is probably not engaged completely, so give it another shot.

Once you're sure the converter is okay, find the converter-to-flexplate bolt hole you marked when the engine was removed. Position it directly at the bottom of the housing so the stud and the flexplate can be aligned when the engine is installed.

MANUAL TRANSMISSION

There's not much to worry about if you have a standard transmission. Sometimes, gear lube will leak into the bellhousing. If you find no evidence of leaks, then leave well enough alone. If you do find transmission grease in the bellhousing, look to the input shaft seal as the culprit.

To replace the seal, remove the input shaft bearing retainer. It comes off with four bolts. Set it on a solid surface, then use a long punch to drive the seal out through the transmission end of the retainer. Work slowly and carefully around the perimeter of the seal.

The new seal is installed the same way as an automatic transmission front seal. Use the old seal as a bumper and remember to install the seal so the lip points toward the transmission. After the seal is in place, wipe a little oil around its lip and reinstall the bearing retainer with a new gasket. Tighten the retainer bolts to 35 ft-lbs.

Clutch Inspection

If your car has a standard transmission, now is your last chance to inspect or recondition the clutch disc, pressure plate and flywheel. To check the disc, get out your micrometer again and check its thickness. You must compress the wave (marcel) spring inside the plate to get an accurate measurement.

Warning! The slightest amount of grease on the face of the disc will cause "chatter"—rapid grabbing and releasing of the clutch. This is an ugly problem but can be prevented simply by being careful. Use new, clean shop towels wrapped around the disc so grease doesn't get on it.

Lightly clamp the disc in a vice, again, with a shop towel between the disc and jaws of the vice, compressing it until the wave spring is flat. Measure the thickness at the point closest to the vise. A new disc measures about 0.330-in.

At about 0.280-in. the rivets are flush with the friction surface and ready to groove the flywheel and pressure plate. Chrysler recommends replacement when the facing is within 0.015-in. of the rivets.

Let's do a little arithmetic. Suppose your clutch disc measures 0.300-in. and it's been in use for 35,000 miles. You've used 0.030-in. to go 35,000. By extrapolation you have about another 30,000 to 35,000 miles left.

Use your own judgment on whether to replace the disc now or later. It sure is easy to replace now. Later, well. . .

Pressure Plate—My approach to pressure plate inspection goes like this: if it was okay when it came out, it'll be okay when it goes back! Pretty simple. But let's refine it just a bit.

If the plate looks good—bright and shiny, free of rivet gouges, hot spots, or heat checks or cracks, it's probably okay. But if it has hot spots (black or blue spots on the face) the plate overheated at one time. It very likely has been overstressed and should be replaced.

Also, if the release levers are badly worn where they contact the release bearing, replace the pressure plate.

No matter what the plate looks like, if it has serviced more than two clutch discs it must be replaced. Chances are the bushings in the release mechanism are worn and the pressure plate springs are fatigued. This reduces clutch capacity. The newly rebuilt engine will develop more torque than the old one, so it'll need the additional clutch capacity.

Flywheel Inspection—If it happens to the pressure plate, it also happens to the flywheel. The flywheel, however, has larger mass and thereby suffers less. Heat that would warp a pressure plate can be absorbed by the flywheel.

The flywheel is not, however, immune to grooving by rivets. Grooving, hot spots and heat checks can be removed by resurfacing—one of the chores for your machine shop. If they can't do the job, they'll be able to direct you to a shop that can.

Be sure the flywheel is ground, not turned on a lathe. Turning on a lathe can produce bumps on the surface where the hot spots have hardened the metal. A ground flywheel surface will be flat.

We've one final problem to deal with: if heat checks have developed into long radial cracks toward the center, the flywheel must be replaced. These are stress cracks. Under the stress of even low rpm, a flywheel like this can literally explode—just like a hand grenade. Save a leg. Don't ignore this warning. If you see cracks, toss the flywheel.

Be sure to clean the flywheel. Grease here will cause chattering just as I described for the disc.

Deposits may be removed with 400-grit sandpaper. This will also give the flywheel a little tooth. Follow up with a thorough cleaning with lacquer thinner, MEK or toluene. Avoid petroleum (oil) based cleaners such as kerosene or cleaning solvent. Treat the pressure plate to the same sanding and bath.

Clutch Release Bearing—Just replace it. Don't fool around with it. Get a new one and be done with it. It's not expensive and it's a heck of a lot easier to replace now than after the engine's installed.

The clutch release bearing presses onto its sleeve, which slides on the nose of the front bearing retainer. To remove the bearing, slide the release bearing sleeve off the transmission and place the sleeve in a vise but don't tighten it. Rest the bearing on the top of the jaws. Carefully drive the sleeve out of the bearing.

To install the new bearing, place it over

The six bolts retaining the flexplate must be torqued. Don't try to do this with the engine still on the stand.

Up and in! Two people make this a simple task. You can do it by yourself but plan on it taking an hour or more if it's your first time.

the sleeve, set the old bearing against the new one. Now clamp the whole works in the vise and use the old bearing to press the sleeve into the new bearing. Finally, fill the groove in the hub with moly grease.

PREPARATION FOR INSTALLATION

You're now ready to get the engine off the stand. You must do this to install the flywheel or flexplate. Just as you did in Chapter 2, use a cross chain or lifting plate with your "cherry picker" and lift the engine enough to take the weight off. If you plan to use a chain bolted across the engine, take this following good advice. Adjust the chain and the hook on your lifting device so the engine drops down just a bit in the rear as it's suspended—maybe 4 or 5 inches. This will help align the engine with the transmission which you will have jacked up a little bit prior to beginning the installation.

If you're using a lifting plate on the carburetor mounting bolts, there will be a natural drop because of the weight differential, and it will be easier to move the engine by hand with this type of arrangement.

Remove the stand from the rear of the engine. Check the crankshaft mounting flange to be sure it is free of nicks and burrs that might cock the flexplate or flywheel.

Position the flexplate or flywheel on the crankshaft, convex side out, and align the six bolt holes. Apply a little Loctite to the threads, then install the six mounting bolts, torquing them to 55 ft-lbs. Note the ear of the flexplate with the paint and rotate the crank so this hole is at the lowest position.

Mounting the Clutch

With the flywheel in place you can install the clutch. Remember, do not touch the friction surface. Before installing the clutch, wipe the disc down again with lacquer thinner. From now on, handle the disc only by the edges.

If you have an old transmission input shaft, it's great for aligning the clutch disc to the pilot bearing. If you don't have one, you can buy a dummy shaft at the parts store for just a few bucks. If it's only going to be this one-shot deal, buy the cheapest.

Slide the disc onto the tool and insert the tool into the pilot bushing. Position the pressure plate against the flywheel and disc, and thread the bolts in finger-tight.

Tighten each bolt a couple of turns at a time, moving around the pressure plate until the cover is firmly against the flywheel. Tightening one pressure plate bolt all the way will bend the cover. Remove the alignment tool after making sure it slides in and out easily. Torque the 5/16-in. bolts to 17 ft-lbs. and the 3/8-in. bolts to 30 ft-lbs.

At this point, the engine is ready to be installed in the car.

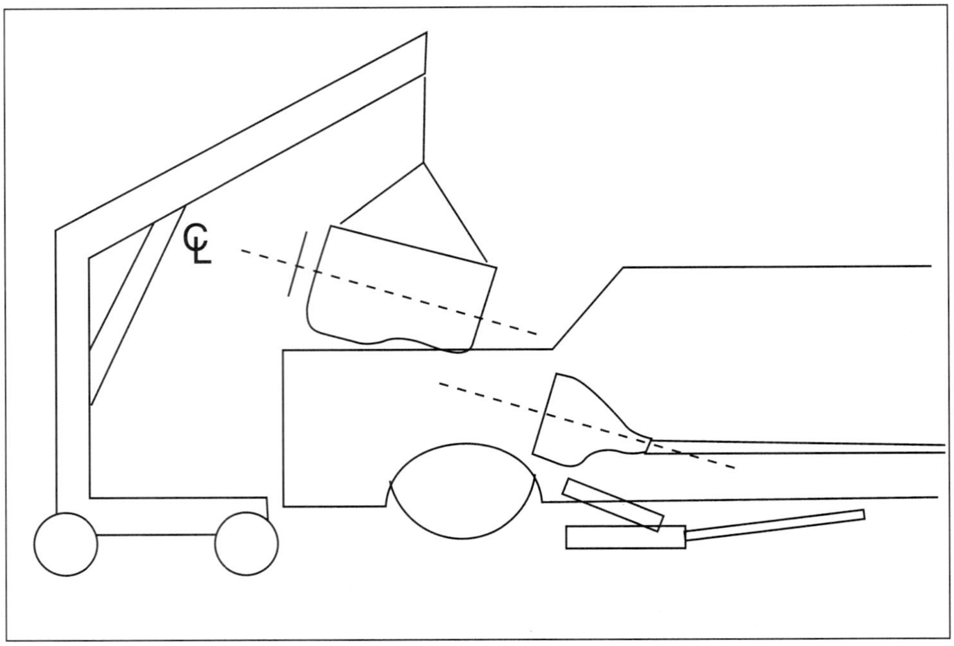

Before lifting the engine to set it in the engine compartment, adjust the lifting chain to allow the back of the engine to tilt down—such that the centerline of the engine matches the centerline of the bellhousing. This will align the contact faces of the engine and bellhousing to the same plan. Getting the first bolts through the bellhousing into the block will be a snap if you follow this procedure.

INSTALLING THE ENGINE

Go over the engine now inch by inch to be sure everything has been installed—especially all the plugs, both oil and water. Is the oil pan drain plug in? Are all the oil gallery plugs installed? If they're not, you'll certainly know it when you start up the engine.

If you're working with an automatic transmission be sure the flexplate ear with the paint dab and the corresponding bolt hole in the converter are at the bottom. If you forget to do this you'll really regret it. Now, let's reunite engine and car once again.

Lifting the Engine

No matter how many times you lift an engine in or out you always have your heart in your mouth wondering if anything is going to let go. If you were careful and did everything right you shouldn't have a problem, but prepare for the unexpected. If you're concerned at all, go back and reread the chapter on removal and double check everything. The safety precautions are the same.

I really suggest you find a friend who can help you here; even one who knows nothing about engines. One of you can do the running around while the other guides the engine into place.

Place your floor jack under the transmission with a block of wood below the transmission pan to prevent denting it. Lift the transmission until it almost touches the firewall—almost but not quite. Be sure you remove the wire you had holding the tranny in place.

Lift the engine over the engine compartment. If you happen to be working in a garage with less than an 8-ft. ceiling you might run into trouble getting the engine high enough to clear the body. A sneaky little trick here is to let some air out of the tires (remove the jack from under the transmission first). This will give you another 3 or 4-in. of clearance!

Carefully guide the engine back, aligning the rear face with the bellhousing. Keep the gap between the rear face and the bellhousing equal from side to side and top to bottom. This is where the rearward tilt you introduced earlier will give you tremendous advantage.

With manual transmissions, put the transmission in gear. When the disc and input-shaft splines contact each other, rock the engine back and forth while holding the engine against the input shaft. If the splines don't engage, turn the crank with a wrench to change the alignment.

Once the splines engage, put the

This is the starter mounting stud on the bellhousing. It has no part in the mating of the bellhousing and engine block. Be careful not to try to line up any of the mounting holes on this stud.

It's in now—seated and ready to be bolted up. Note the shop towels in the carburetor and distributor ports. This keeps your buddies from dropping orange peels in here.

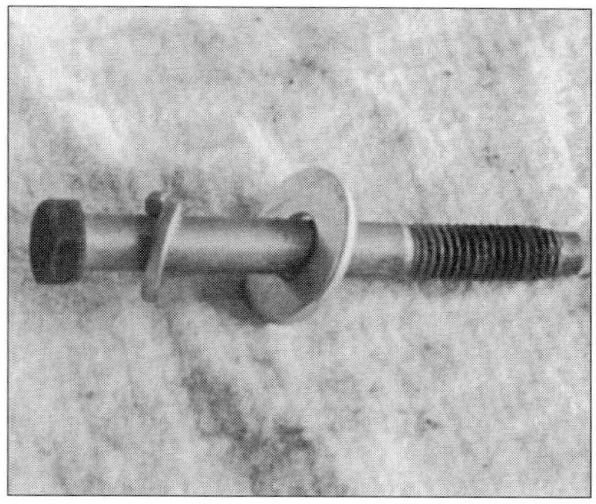

This is the through-bolt for the engine mount. Be sure you have a flat washer and a lock washer on it. The passenger side enters front-to-rear and the driver's side enters from the rear-to-front.

transmission into neutral to allow the engine to rotate into position with the bellhousing dowels. When the engine and bellhousing come together, they'll "snap" together.

If your car has an automatic transmission, it will work almost the same way. However, there's no help from the transmission input shaft. You'll have to get into the engine compartment and guide the engine with hands, feet, knees or whatever. Here's where that friend can help by raising and lowering the transmission to align it with the face of the block while you jockey it into place.

When the block and converter housing or bellhousing come together, get a couple of bolts in to hold them together. Move to the front of the engine and look at the engine mounting brackets. More than likely they won't be aligned with the mounting bosses on the frame. In fact, it's likely that one of the "ears" of the engine mount will be stuck in the slot in the center of the boss. You'll have to do some prying with a big screwdriver or pry bar here to get this out.

If the pry bar doesn't work, remove the jack from the tranny, raise the engine and shove everything back an inch. When both engine mounts are aligned over their bosses, let the engine all the way down. The engine is in and won't go anywhere. Before you remove the lift or jack, get all the bellhousing bolts you can access from topside in and tightened. Now you can remove the lift and the jack and nothing will happen.

The passenger side engine mount can be reached from the top. Get out one of the two engine mount through-bolts and run it in. Remember that the nut is already welded to the mount. Don't spend all afternoon looking for it. Access the driver's side from beneath the car after you jack it up.

Jack up the front of the car and get some safety stands under it. When the car is secure on the stands, rock it back and forth to see if it really is stable. Better to push it off the stands while you're outside rather than when you're underneath. I chose to install my engine by working from the bottom up, but you can do it any way you like.

Always, always place your car on safety stands. Then, rock your car by hand to check that they are sturdy and stable. The car shouldn't move at all in order to be safe.

Bellhousing to Engine

Install the remainder of the bellhousing bolts from under the car. The tranny dipstick tube attaches to the bottom bolt. Torque all the bolts to 35 ft-lbs.

Flexplate to Converter

Install the first bolt through the flexplate into the converter and tighten it to 25 ft-lbs. Use a wrench on the damper bolt to rotate the converter and flexplate clockwise as far as you can. When you can rotate it no further, pry on the ring-gear teeth with a large screwdriver. Install the second bolt and tighten it to 25 ft-lbs. Install the remaining two bolts in the same manner. Install the converter cover and tighten these bolts to 10 ft-lbs.

Starter Motor

Unlike most cars, the starter for the big-block Mopar goes in fairly easy. Hoist it up and into place. Secure it with a nut over the stud in the bellhousing. To tighten the nut you might want to use an 18 or 20-in. extension on your ratchet. This gets you back where you can get some room to operate. The top bolt can be accessed from beneath the car but it's easier to reach from topside. Torque the nut and the bolt to 50 ft-lbs., then connect the starter cables.

Driver Side Engine Mount

Swing around from the starter motor and install the engine mount through-bolt. Torque to 75 ft-lbs.

Exhaust Pipes

Again, you'll be well ahead if you can find a friend to help you with the exhaust pipe lace-up. The exhaust pipe mounts to the exhaust manifold with nuts and bolts. If there's one person on top and another under the car, the job goes quickly and effortlessly. If you're installing the original nuts and bolts they'll be rusty. Save yourself some grief and run a die down the bolt and a tap through the nut. They'll go together so much easier and nicer.

You must have a gasket between the pipe and manifold flanges. If it's missing from your kit, they're readily available at the parts house.

Push the exhaust pipe up to the manifold with the gasket in place. Have your partner drop a bolt through while you work a nut on to it from below. You will both be more comfortable with ratchets and long extensions. Torque the bolts to 40 ft-lbs.

This completes everything you need to do under the car. Remove the jack stands now and set the car level on the floor. Let's get ready to do the top side.

TOP SIDE

If your car is equipped with an air conditioner, move directly to that part of the chapter, page 155 under *Air Conditioning Compressor* and begin with the installation of the air conditioner and the various accessories. Because of its size and many mounting points, it's easier to do it first then work the other parts in around it. If you don't have A/C, start here with the carburetor. I've begun with the carburetor for two reasons. First, under most conditions it's the proper place to begin. Second, it's easier for you to see the various mounting points without that huge pump in the way.

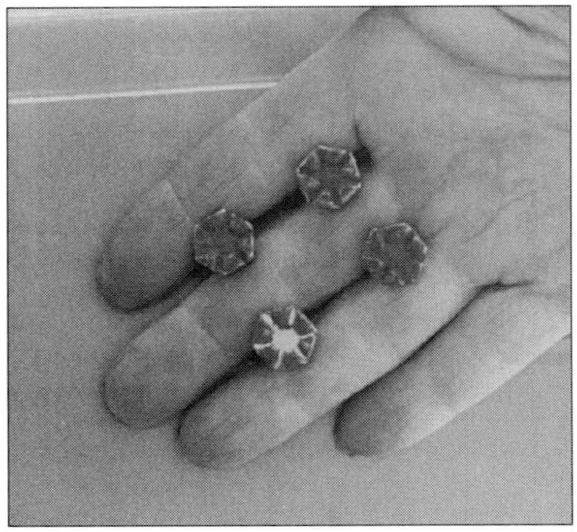

These are the four mounting bolts for the flexplate to torque converter. Notice the bottom one still has paint in its grooves from when I marked the flexplate ear with paint.

Fuel Line & Filter

If you have any concerns about the condition of the fuel filter you should replace it. Most filters will include new rubber tubing and spring clamps, but if not, get two pieces of 5/16-in. rubber fuel line, not vacuum hose, and replace the old tubing.

Reach down between the block and fender apron and find the fuel line. Remember, you stuck a bolt in it to keep the fuel from draining out. Remove the bolt and fasten the line to the fuel pump.

The flexplate and torque converter are "married" again. I'll replace the cover next. Notice that I aligned plate and converter at their original location.

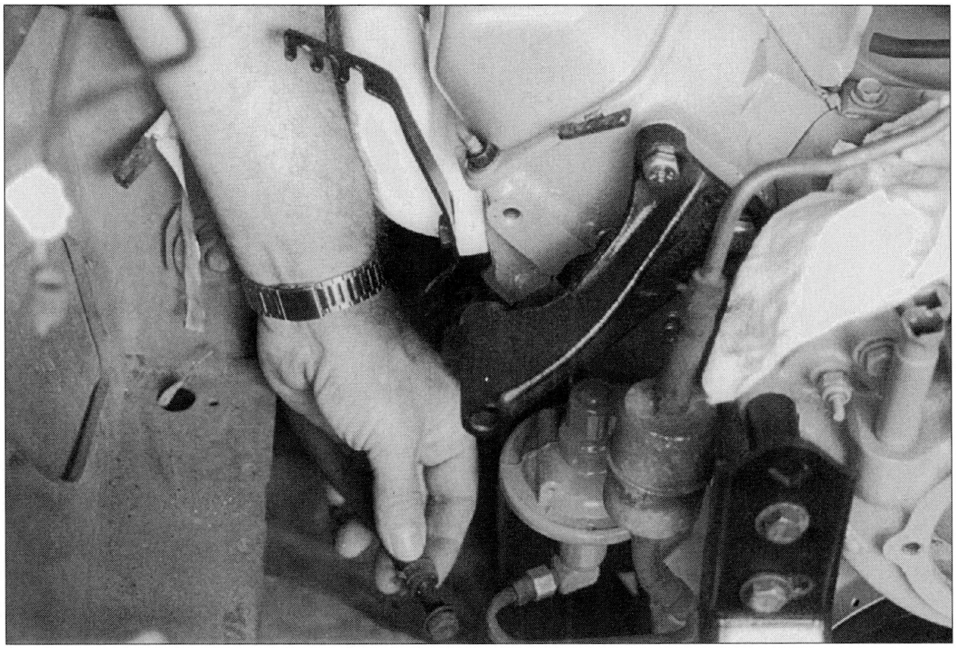

If you forget to hook up this fuel line you'll never get the engine started.

I preach the use of flare-nut wrenches to save you time and money. If you round off the corners of this nut before it's tight, the only way you'll stop the ensuing fuel leak will be with a whole new fuel line and flare nut.

Carburetor

Locate the carburetor gasket that you should have ordered along with all other gaskets. As I mentioned in Chapter 7, most gasket kits won't have one for the carburetor; it must be ordered separately. Once you find it, mount it in the correct direction. It will be stamped front or rear. If you install it backward you'll incorrectly divert the flow of fuel.

Mount the carburetor and torque the mounting nuts and bolts to 15 ft-lbs. This is an important number. Exceeding 15 ft-lbs. of torque can crack the base of the carburetor or strip the threads of the studs or bolts.

Install the fuel line. You took this line off with a flare-nut wrench to prevent rounding the corners of the connector nut. Use this wrench again, and on all other flare nuts to prevent rounding the corners.

Hook the choke heater to the choke to avoid forgetting it. The end is retained with a clip. Now move to the other side and connect the throttle linkage.

Begin with the automatic transmission kickdown lever. It's stuck down there between the engine block and the firewall. Locate it, lift it up and snap it onto the bellcrank pivot ball. Locate the long, adjustable rod. It runs between the top of the bellcrank and the outboard side of the throttle lever. A clip retains it on the throttle lever and the long, thin throttle spring retains it to the bellcrank. Don't forget the washer that fits between the spring hook and bellcrank. The other end of this spring hooks in a slot in the bottom of the dogleg at the outboard end of the throttle lever. Find the short, fat spring and connect it from the cable attachment at the throttle lever to the throttle-spring bracket at the front of the intake manifold.

Hopefully, it will all fall into place as you assemble it. Keep checking with the photos nearby to be sure everything is going together correctly.

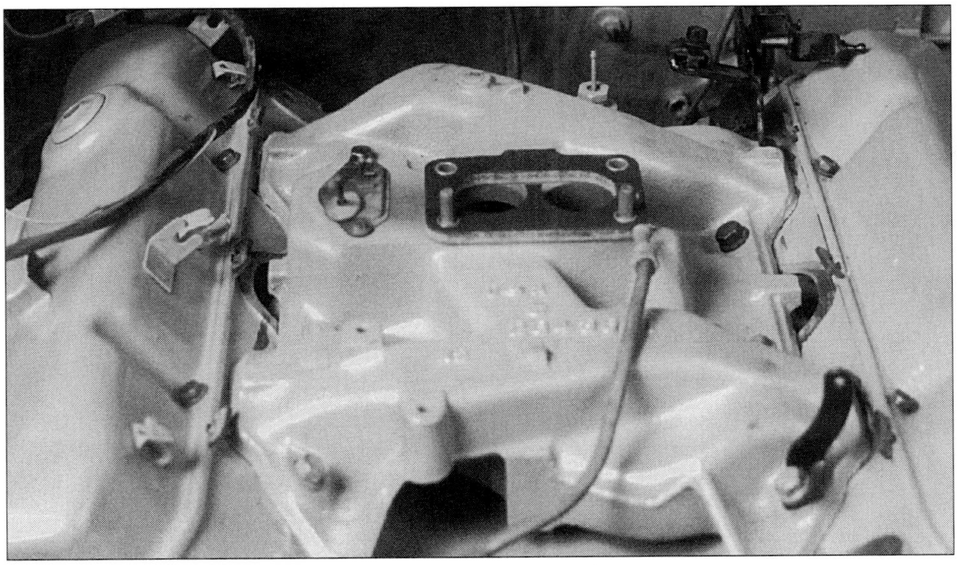

Be certain that your carburetor gasket is facing in the right direction. It will be marked front or rear for your convenience.

PCV

To check the old PCV valve, shake it. If it rattles, it's okay. If not, soak it in solvent for a while. If it won't loosen up buy a new one. Without it, the crankcase won't vent properly and you'll never pass your smog test.

If the rubber hose is cracked or weak, replace it. 3/8-in. vacuum hose is fine.

Spread a little petroleum jelly on the PCV valve and insert it into the rubber

Here's another of those little things that are easy to forget—the choke heater rod.

grommet on the passenger side valve cover.

Distributor & Spark Plug Wires

To prevent installing the distributor 180° out of time, follow the directions below very carefully.

Select the correct O-ring from your gasket kit and install it on the distributor drive. This will prevent future oil leaks. Lubricate the O-ring with petroleum jelly.

Remove the distributor cap. Be sure the rotor is in place. Hold the distributor over its mounting hole in the manifold with the vacuum diaphragm pointing forward (away from you and toward the driver side fender). Point the rotor to the 4:30 position (as if the rotor were the small hand on a clock). Check to be sure the rectangular shank at the end of the distributor drive is parallel to the centerline of the cam and would drop into the slot in the oil pump (distributor) drive gear. The rotor is now pointing to the number 1 position in the firing order. Make a mark on the housing corresponding to the direction the rotor is pointing. When you replace the distributor cap, the tower (socket) above the mark on the housing will be the number 1 spark plug terminal.

The next step is to determine if number 1 piston is at TDC on the firing (compression) stroke. You've no doubt turned the crankshaft since installing the distributor (oil pump) drive gear. So, refer back to page 135 and refresh your memory on how to find TDC on the number 1 cylinder.

With piston number 1 at TDC and the distributor cap removed, line the rotor up with the mark on the housing and gently insert the distributor into the engine. You may need to wiggle the rotor right or left a bit to seat the shaft in the drive. The distributor is seated when the shoulder on the housing is flush with the block.

Install the distributor clamp and snug the bolt for now. You'll loosen it later to adjust the timing with the engine running, but the engine should fire with this timing.

Now, if you forgot to number the plug wires or plan to use new ones, here's how to lace them up. First, make a mental note to yourself that rotation of the distributor is counterclockwise. The firing order is 1-8-4-3-6-5-7-2 and the cylinders are numbered 1-3-5-7 on the driver's side and 2-4-6-8 on the passenger's.

Begin with number 1 on the distributor

I hope this view of the carburetor linkage can help you get everything hooked up right. It's easier to see this way than after we get the A/C unit in.

Any time you have to mate rubber and metal together by forcing one over the other, lubricate one or both surfaces with petroleum jelly. Al is getting ready to insert the PCV into the rubber grommet in the passenger side valve cover.

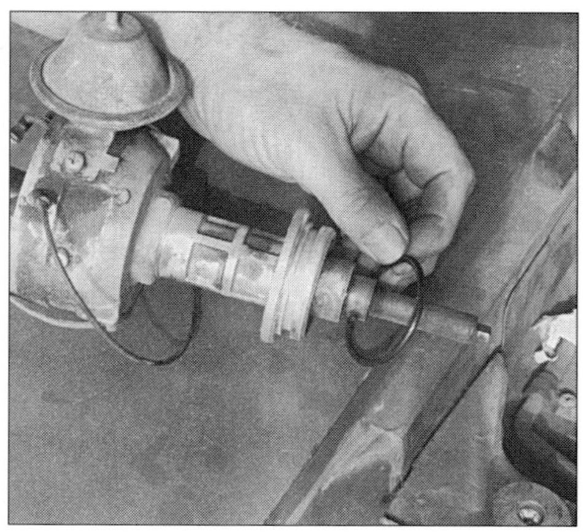

A nice, new "O" ring for the distributor. It's in your gasket kit.

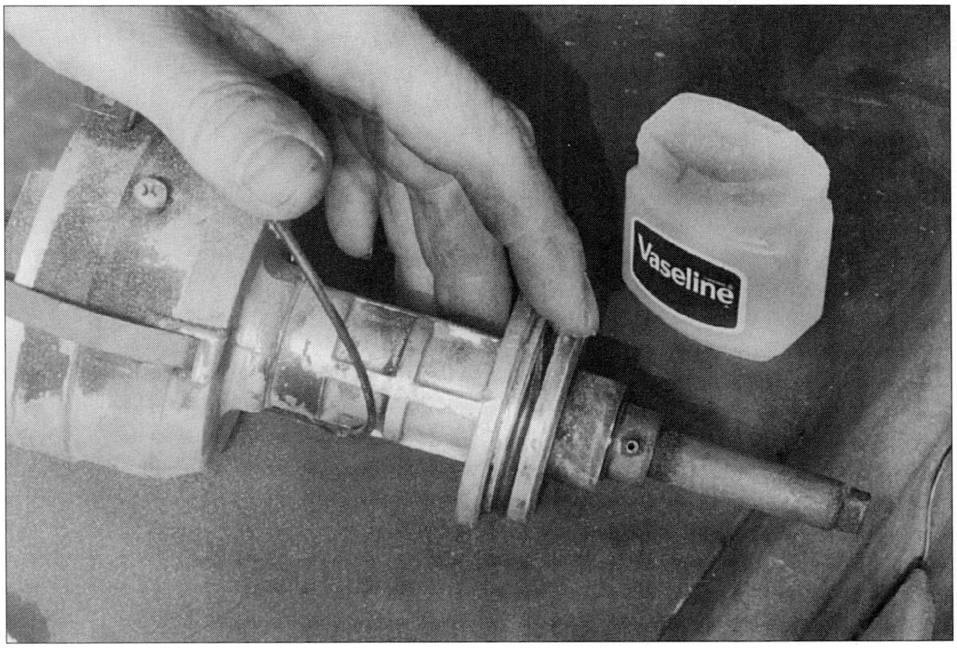

As you did with the PCV valve, lubricate the rubber to metal surface with petroleum jelly.

cap and run this wire to the first plug on the left bank. Moving counterclockwise around the distributor cap, the next wire goes to number 8, then number 4 and so on through the firing order.

Don't forget to hook up the vacuum advance hose from the carburetor!

Heater Hoses

If you have no air conditioning, install the heater hoses now. If you're fortunate to have air, wait until you've got the A/C bolted in before you do the heater hoses.

The heater hose coming from the right side of the heater core goes to the rearmost nipple. The hose from the left line goes to the forward nipple. A little silicone sealer goes a long way toward preventing leaks. Don't forget to tighten the clamps.

Air Conditioning Compressor

There are 8 attaching points for the A/C compressor. We'll bolt in 5 now and 3 later. Remove the right-front intake manifold bolt, and the first and second left intake manifold bolts. Remove the front most exhaust manifold nut.

Set the compressor in place, being sure the bracket to the exhaust manifold stud is in place. Install a bolt in each of the top mounting points and a nut over the exhaust manifold stud. For now, these should only be finger-tight. Leave the support bracket from the front mounting boss on the intake manifold open for now. Later, you'll install the coil here.

Power Steering Pump

The power steering pump mounts to the A/C compressor on its two mounting bosses to the left. Use 2, 3/8 x 3-1/2-in. bolts. They pass all the way through the mounting bracket and into the water pump housing. These also should not be torqued yet. Wait until it's time to install the belts.

Water Pump & Fan

If you removed the water pump and fan as a unit when you disassembled the engine, you'll be ahead in the reassembly. In fact, it's best to join these two components if you've replaced either one. It's much easier to install them as a unit than as separate pieces.

Cement the gasket to the pump. Then run another bead of silicone sealer around the mating surface to the block. Installing the pump is just a matter of installing the four bolts and torquing them to 30 ft-lbs. However, be sure the weep hole in the pump is pointing down. Omit this step and the pump will soon rust out.

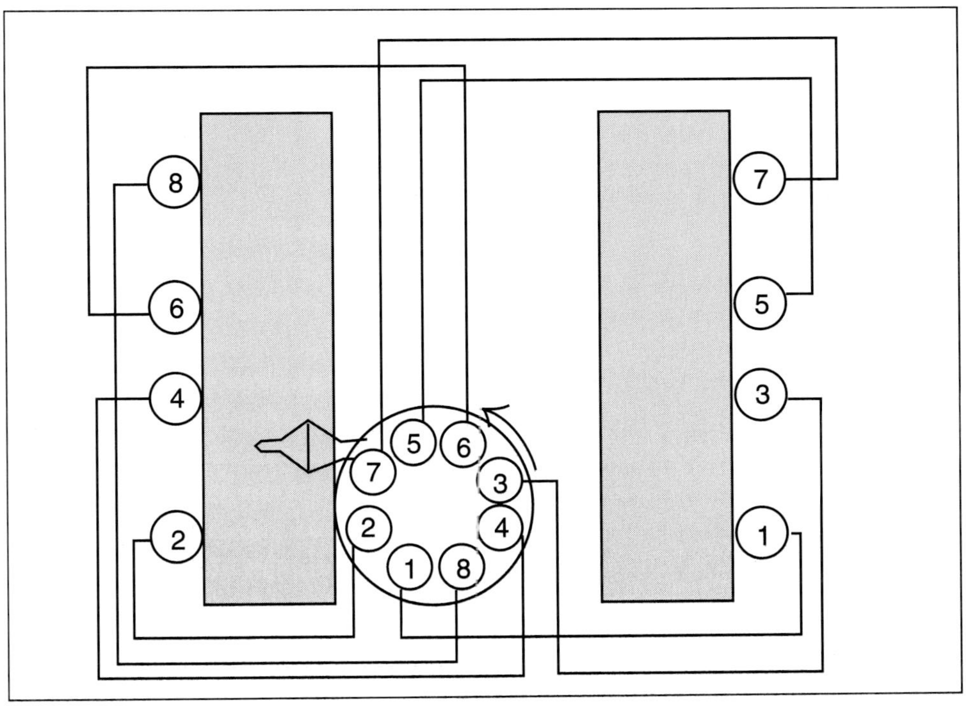

Follow this schematic for the correct route of the spark plug wires. This is a schematic only and not the way for you to run the wires. The wires are all run through the looms on the front and sides of the block—except for wire #8 which travels from the distributor, along the valley side of the right valve cover, around the back of the right head to the plug.

Why is the left valve cover off? Correct! I have to find TDC on the compression stroke to time the distributor. If I used TDC on the exhaust stroke I'd be 180° off. I had to find TDC again after Al turned the crankshaft a couple of times looking for the flexplate ear with the paint on it.

Alternator

The last big "to-do" on the front end is the alternator. If you left the long support arm on, it will fall right into place. If you removed the support arm it should be installed so the heater hose clamp bracket is in the up position and the two 1/2-in. holes pointing toward the A/C bracket to which it mounts. The heater hose support bracket mounts to the bolt attaching the other end of the support arm to the alternator. The loop of the hose support bracket faces the firewall.

Mount the alternator to the alternator mounting bracket on the right head. Anchor the support bracket through the inboard hole securing it through the A/C brace and into the water pump housing. Leave everything loose as above. You'll torque it all down soon.

Coil

The coil is the last accessory to be installed. It's last because one of its legs slides under the A/C support bracket. This is the rearmost "leg" of the compressor. Slide the distributor mounting bracket under this leg with the lead end of the coil pointing to the right side of the car. Install the two mounting bolts. The condenser goes under the forward bolt—the one retaining the compressor support bracket.

Now, everybody is attached. It's time to torque the bolts for the air conditioner.

Torque Sequences—Tighten everything "wrench-tight" except the adjusting bolts on the power steering pump, the water pump idler pulley and the alternator. Torque all the A/C-to-engine block bolts to 30 ft-lbs., including the support bracket over the coil. Torque the other coil bolt to 15 ft-lbs. The exhaust manifold stud-nut should be torqued to 30 ft-lbs. Torque the A/C lower mounting bracket bolt (the one anchoring the alternator support) to 30 ft-lbs. Torque the water pump to water pump housing bolts to 30 ft-lbs. (Chrysler loves 30 ft-lbs.!)

Belts

You are now blessed with four belts to lace up. They'll only go on in one sequence. Here it is: The two longest belts drive the A/C compressor and the alternator. Install these first. Pull the alternator in toward the center of the engine as far as it will go. Install the two belts around the alternator, over the A/C compressor and down to the crank or harmonic balancer pulley. Snug up the alternator finger-tight so the belts don't slip off.

Here's what the front of your engine will look like if you have no air conditioner. Notice the careful lacing of the spark plug wires through the wire harness. This is more for appearance than performance.

The third (and smallest) belt connects the harmonic balancer pulley to the power steering pump.

The fourth and final belt passes around the water pump, down to the harmonic balancer pulley and back up around the water pump idler pulley. Now you can tighten everything.

Tightening Sequence— Start with the power steering pump. Get out your 1/2-in. breaker bar and your ratchet with an 8-in. extension. Insert the socket drive of the breaker bar in the 1/2-in. square hole in the top of the power steering pump. Pull up on the pump until the belt can only be deflected about 1/2-in. then secure the lock bolt. More than 1/2-in. deflection in the belt can lead to slippage. If the belt is too tight it causes excessive side loading in the P/S pump, resulting in bearing and seal wear. Torque both the retaining bolt and lock bolt to 30 ft-lbs.

The same process is used to tighten the water pump belt. The idler pulley has the same square hole for your breaker bar as the P/S pump. Follow the same procedures and torque limits to tighten the water pump belt.

The designer of the alternator neglected to figure in a location for the breaker bar. Therefore, you're on your own for finding a way to put some serious stress on the alternator. Brother Al, in the photo on page 159, is using the engine turning handle from the engine stand as his pry bar. He pulls back on the alternator and anchors it with 15 ft-lbs of torque. (Note the "torque wrench" forearm. As I said, this guy knows his stuff!) Belt deflection should be about 1-in. with about 15 to 20-lbs. pressure.

All of the accessories are now mounted, belted, adjusted and torqued. All that remains is to connect the A/C and P/S hoses.

A/C & P/S Hoses

Mount the steel suction line from the A/C, then install a new O-ring on the male end of the A/C hose and lubricate it with refrigerant oil—not motor oil. Insert the hose into the coupler and tighten the retaining nut to about 20 ft-lbs. You'll have to have the A/C recharged with freon. This is no longer a do-it-yourself project and must be done by an EPA-certified establishment. Don't forget to do it or you'll be sorry come summer!

Connect the power steering lines. During all the "taking off" and "putting on" you've lost some P/S fluid. Be sure to check on this before you take the car out for a spin.

This concludes the installation of the

These are the main mounting points for the A/C. Get them on here and the manifold and front mounts will fall into place.

The power steering pump mounts to the bottom of the A/C. The mounting bolts pass all the way through the P/S pump, the A/C and the water pump housing before finally threading into the block.

Mount the water pump and fan blade clutch as one unit. Don't forget to seal the gaskets with silicone sealer.

Finger-tighten all the bolts as you install the accessories. Then, make them wrench-tight before you start torquing. The bolt Al is working on must be tightened now, before he installs the coil as this bolt is directly beneath it. Torque to 30 ft-lbs.

accessories. Now go back to page 152 and install the carburetor, fuel lines, distributor cap, spark plug wires and the two heater hoses. When you're done, come back here and read the directions for wiring everything back up.

Electrical Wiring

Let's start at the rear and work forward. The first thing to catch is the ground strap from the engine to the firewall. This grounds the two so the instruments have a good, solid ground connection. Without this you can get some real crazy instrument readings. This wire goes from the right side of the block to the firewall.

Coil—The condenser and ignition wires go to the terminal marked BAT or +. The ignition wire is the blue one from the main wiring harness. Next, connect the primary lead from the distributor to the other terminal on the coil. Then install the "hot wire" from the center of the distributor to the coil.

The best theft insurance, other than a Rottweiler in the front seat, is to simply remove this wire when you park the car. Can you imagine a thief trying to cobble something together to take its place?

Alternator—On single-field alternators you will find two wires, one green with a bayonet clip and one black with a loop or ring connector. The green wire goes to the terminal marked FLD and the black wire to the terminal marked BAT.

Dual-field alternators have three wires. The loop connector hooks to the BAT terminal while the two bayonet connectors go to the field terminals. Looking at the back of the alternator with the battery connector at 12 o'clock, the blue or blue-and-white wire connects to the field terminal at 9 o'clock. The green wire connects to the terminal at 11 o'clock.

Cars equipped with a 100-amp alternator have an additional, external ground wire. This plainly marked wire on the back of the alternator bolts to the head. Both field wires use a single plug-in connector. The battery wire is a loop connector.

The last electrical hookup is the battery ground. This connects to the inside front of the left cylinder head. Check the cable—and the "hot" cable, while you're at it—for corrosion at the battery connector. If either is badly corroded replace the cable or cut the connector off and install a universal clamp-on connector. Clean the inside of both clamps with a knife or battery brush. Install the battery cables.

Fan Shroud & Radiator

Unless you recently replaced the radiator hoses, do yourself a favor and install new ones.

If the ends are not flared or ragged and you think there is still a lot of mileage left

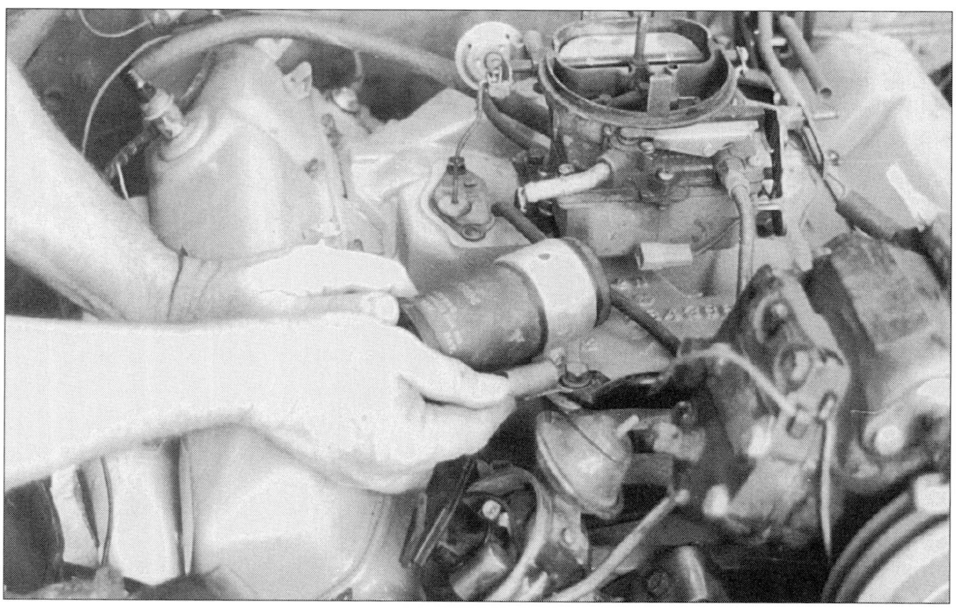

You'll need this condenser to improve your radio reception. Without it you get no "tunes."

Both the P/S pump and the water pump idler pulley have a 1/2-in. cutout for your breaker bar. Use it to adjust belt tension. Then tighten the securing bolts to 30 ft-lbs.

in the old ones, try this test: squeeze each hose flat. If the hose is firm with no cracks along the pinched edge, the hose is probably okay. Pay particular attention to the lower hose. It runs hotter, so it's the first to fail—and tough to replace.

If one radiator hose needs replacing, replace both. This prevents the hassle of replacing one hose now and having the other one fail a month later.

The fan shroud is next. It seems logical to install the radiator before the shroud. It won't work; the shroud won't slip between the radiator and the fan. Hang the shroud back over the fan, install the radiator and tighten the six radiator-mounting bolts.

If your car has an automatic transmission, there'll be two cooling lines at the bottom of the radiator for the ATF. The nuts on these lines are brass, so be sure you use a flare-nut wrench to prevent rounding the edges.

Al resorted to the bar with which you rotate the engine in the engine stand to bring tension to the alternator belts. You can use pipe, a crowbar, or a 2x4. Broom handles usually break!

While you have your head down there install the bottom radiator hose. Be careful not to over-tighten the clamp at the radiator. The neck is thin brass and will collapse with little effort.

Now you can bolt the shroud to the radiator. Use a washer between each bolt and the shroud to spread the load and prevent cutting the plastic. Tighten the shroud bolts about 8 to 10 ft-lbs. Any tighter can damage the plastic. Finally, install the top radiator hose.

Inspect the engine compartment carefully. Make sure everything is connected and remove any loose tools or parts. You're about to fire the engine for the first time in its new life.

ENGINE BREAK-IN

Before starting your newly rebuilt engine keep the following three points in mind. These will ensure the engine will provide thousands of trouble-free miles.

Lubrication

Throughout the assembly chapter I stressed cleanliness and lubrication more than anything else. This was done to provide immediate lubrication to all engine parts.

To be sure the engine develops full oil

Another O-ring seals the A/C line.

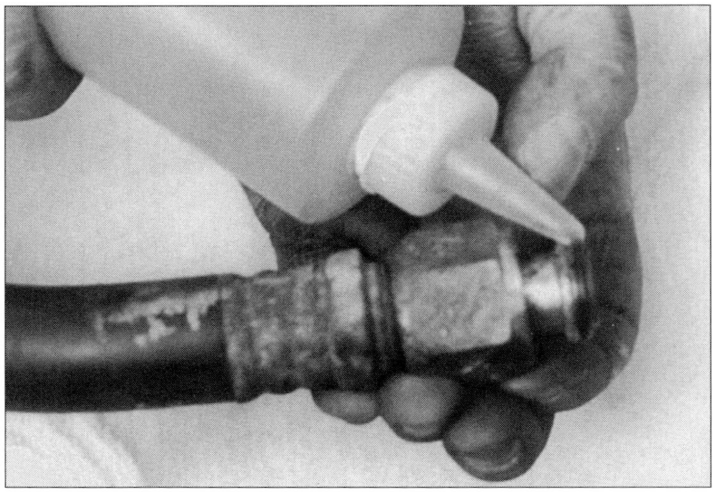

Be sure you lubricate this O-ring with refrigerant oil. In this case do not use petroleum jelly or motor oil.

pressure, remove the spark plugs so the engine will crank faster; then crank the engine. The oil-pressure idiot-light should go out within 15 seconds. A gauge should indicate 35-40 psi within the same time. Once the oil pressure is okay, turn your attention to engine cooling.

Cooling

The additional friction of a new, "tight" engine produces more heat than after it has a few hundred miles on it. Also, less air passes through the radiator simply because the car is not moving. The additional heat and lost cooling efficiency add up to the makings of an overheated engine.

To keep a constant flow of cool water through the cooling system, stick a garden hose into the radiator and open the radiator petcock. Fill the radiator and engine, then adjust the flow into the radiator to compensate for the flow out. Readjust the flow once the engine starts. This will keep up a constant flow of cool water to the engine and prevent overheating.

Finally, hook a timing light up to the number 1 plug wire. Loosen the distributor clamp, and disconnect and plug the vacuum advance hose.

If your car has an automatic transmission, be ready to top it up with fluid. You drained the torque converter, so the transmission is going to be down about 5 quarts.

Start-Up

When you fire the engine, run it for 30 minutes at about 2000 rpm. Watch your gauges to check for overheating or loss of oil pressure. If either occurs, shut off immediately. This assures a constant flow of oil to the cam and lifters. If the engine is idled at a lower rpm, insufficient lubrication to the cam and lifters could possibly gall one or more lobes— which puts you back to square one.

So, with these three points in mind, let's start the engine.

Light It Up!

When you cranked the engine to check for oil pressure, the carburetor float bowl should have filled. If the engine doesn't jump to life on the first try, splash—don't pour—a little gasoline into the carburetor.

When the engine starts it will blow a huge cloud of smoke out the exhaust. This should stop in just a minute, so don't worry. Also, hydraulic lifters will rattle like crazy until they pump up, or pressurize, but this should only take a few minutes.

Make sure the fast-idle cam on the carburetor is on the fast-idle position and let the engine run for 30 minutes at 2000 to 2500 rpm.

Have your timing light hooked up and ready to go. Timing is set soon after the engine fires.

As soon as the engine starts, check the oil light or gauge, then fill the automatic transmission to prevent any damage.

Once you've taken care of these items, set the timing. Be sure the vacuum-advance line is plugged. The engine may be idling a bit faster than it should but at least the timing won't be so far retarded the engine will overheat. Advance or retard the distributor to line up the timing mark on the damper with the correct timing index on the front cover.

Be very careful around the fan and accessory belt drives. They have the ability to reach out and "bite" you.

Post Break-In Checks

When the engine has run a full 30 minutes, shut it off. Remove the garden hose and drain the radiator. Close the petcock and refill the radiator with antifreeze and water in the ratio specified for your area and season. Never, never run 100% water in a cooling system. Antifreeze is a more efficient coolant than water. It prevents rust and helps lubricate the water pump.

As you're filling the coolant, turn on the heater so coolant will fill the heater core and hoses. Check the coolant level, top off and install a new radiator cap with the correct pressure rating.

Go around the engine to check for leaks, seepage or anything out of the ordinary. When this is done, go back and retighten the hose clamps. Gaskets and hoses tend to creep or relax when they're compressed and heated.

Also retighten the intake manifold bolts and check the exhaust manifold bolts. Once you've done that, restart the engine but idle it at 500 to 600 rpm and recheck the timing.

Air Cleaner

Install the air cleaner before you drive the car. Never drive a car without an air cleaner. Also be sure there's a seal on the carburetor before installing the air cleaner. Without it the carburetor will draw dust and dirt between the top of the carb and bottom of the air cleaner.

Install a new paper filter in the air cleaner, a new seal under the nut on the top plate and secure the air cleaner to the carburetor.

Hood

Find your hood partner and get him to help you put the hood back on. Thread the bolts in just to the point of being snug. Adjust the hood to the locating marks you drew, or to the holes you drilled. Before you slam the hood down and walk off, carefully lower it to check its alignment. Then secure it.

TRIAL RUN

Okay, this is it—the moment of truth. Check the engine compartment again for anything laying around. It's bad form and expensive to scatter tools and parts along the highways and byways. Check under the car too. Running over a jack stand or creeper makes you look like an amateur.

Now, out onto the highway. You can take the car up to highway speeds, but no hard acceleration.

Cruise around and listen for any bumps, grinds, whistles or groans. As soon as the engine is up to normal operating temperature, scoot back home to check things under the hood. Look for leaks, check all fluid levels and give it a thorough going over. Then re-torque the cylinder heads, intake and exhaust manifolds again.

Unless your engine has adjustable rocker arms and you wish to perform a hot lash adjustment, you can now congratulate yourself for a job well done!

Hot Lash Adjustment

In most instances the cold setting you made in Chapter 7 will be sufficient to give you maximum performance. There may be occasion, however, when you may want to adjust the rocker arms with the engine at operating temperature.

This job is hot and messy, but can be made somewhat easier by using rocker-arm clips to prevent hot oil from splashing all over and an 0.027-in. and 0.030-in. go-no-go feeler gauge.

Rocker-arm clips fit over the pushrod end of the rocker arms and prevent oil from being thrown all over the beautiful new engine and freshly cleaned engine compartment—not to mention all over you. Rocker-arm clips are available at most parts stores.

The go-no-go feeler gauge is a lot easier to use than the single-thickness feeler gauge. It is basically a stepped gauge that should slide freely between the valve and rocker arm on the smaller thickness but not fit at all on the larger one when you've obtained the correct clearances.

It makes no difference which valves you set first, intakes or exhausts. Just make sure you get them all. On each head the order for the valves (E = Exhaust, I = Intake) in either direction is: EIIEEIIE.

Intakes are set to 0.027-in. and the exhaust to 0.030-in. Exhaust valves need a bit more time on their seats to cool, so they have additional clearance.

Remember, the valves are going to be opening and closing, so the feeler gauge will be trapped when the lifter is not on the cam lobe base circle.

After going through all the valves on one cylinder bank, shut the engine off and remove the clips. Reinstall the valve cover, wipe up any oil and repeat the process for the other side.

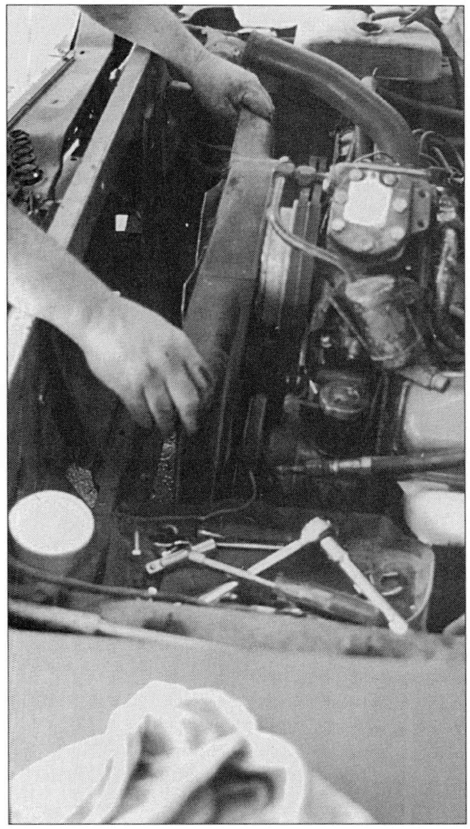

Shroud first—always shroud first. It won't go on after the radiator is in.

Now you can slide the radiator in, attach its six mounting bolts, then pull the shroud forward and mount it to the radiator.

Pour in the remaining 5 quarts of oil. A new engine will suck up almost 6 quarts of oil even though specs call for 5. Why? Because there is no oil film on all the engine parts. You oiled only the moving parts.

After the engine has run for 30 minutes or so, you can then add a mixture of antifreeze and water prior to taking it out on the street.

SUMMARY

Stay alert during the first 200 miles for strange noises and reduced fluid levels—particularly oil level. Check this frequently. Don't be worried if the engine uses a little oil during break-in, especially if you installed chrome rings. Also keep an eye on the oil pressure and engine temperature.

As the engine accumulates miles the idle speed will increase because of lowering internal friction. Don't be surprised if you have to reset the idle several times. Change the oil and filter after 500 miles, when everything should be all settled down.

Make the first miles easy ones. Alter your speed and don't accelerate rapidly.

IGNITION TIMING

A lot has changed in the generation since the introduction of the "B" series engine. Lead is gone from our gasolines; octanes are much lower now. Emission controls regulate our cars' performance. Ignition timing bears the brunt of these changes. I would like to give you a chart here that tells you exactly where to set ignition timing but that's no longer fully possible. As soon as you take your car in for its "smog" (emissions) test, they'll set the timing not for best performance but for lowest emissions.

Below I give you Chrysler's recommendation for ignition timing for pre-catalytic converter autos. Understand, however, that these do not take into account either gasoline quality or altitude. If your car has a catalytic converter you must check your Vehicle Emission Control Information label.

After the car is running and you've put some miles on it, take it to a professional tune-up shop where they can hook up their analyzing machine to the engine and set the timing for all of today's variables.

Chrysler Recommendations:

All engines rough timing = 10° B.T.D.C.

Fine timing 383, 400, 413, 440	
Without C.A.P.	12-1/2° B.T.D.C.
With C.A.P (manual)	Set at T.D.C.
(automatic)	5° B.T.D.C.
Fine Timing 426 Hemi	12-1/2° B.T.D.C.
With C.A.P.	N.A.

Don't do extended freeway driving without changing speed: slow down a little, speed up a little.

Follow these break-in rules carefully and the new engine will show its appreciation with 100,000 miles of excellent service. After 200 to 300 miles on the engine, take your car, van or truck to a professional tune-up shop—one that specializes in tune-ups—and have the engine expertly tuned. See the sidebar below for tips. For now, let me say congratulations on a job well done! ■

It's a runner, folks!

TUNE-UP TIPS

After 200 to 300 miles of street and highway driving, it's time to give your new engine a professional tune-up. Even though you're quite able to tune your own engine, I suggest a professional because a professional tune-up will go far beyond what you're able to do at home. At best you've already set dwell, timing, checked point resistance and adjusted the carburetor. However, most home garages are not equipped with electronic engine analyzers. With the analyzer you can be sure your engine will perform at least as well as when it was new.

If you modified your engine, changed the carburetion or distributor, the engine must be recalibrated to obtain maximum performance. Here the shop will use a very sophisticated piece of equipment called a chassis dynamometer.

The rear wheels of the car drive a set of rollers in the floor of the garage. The rollers can be adjusted to simulate any driving condition. This allows the tune-up mechanic to "see" how the engine performs under most conditions. Additionally, the dynamometer can be used to determine the power at the drive wheels, which means an engine can be tuned for maximum power and mileage and minimum emissions. Be sure your mechanic tunes the engine along the entire performance range rather than for full power only.

Today's emission standards, however, will take precedence over performance. While you're at the tune-up shop they will do all they can to secure optimum performance and keep the emissions within range. If you suggest otherwise, when it comes time for your vehicle emissions testing you'll have to pay extra. So, why not get it all done at once and save a couple of bucks?

The cost of a professional tune-up will be between $50 and $150. It will be repaid in power, mileage and lower emissions—not to mention maximum driving pleasure from your newly rebuilt big-block Mopar.

Martin and Virginia Ahlgrim brought their fabulous 1970 Hemi 'Cuda to Dyno-Tune of Covina, California, for a professional tune-up on a chassis dynamometer. This will insure top performance for this spectacular car. Kevin Iha, Martin Collings and Rodney Lewis of Dyno-Tune stayed long after the shop closed to do a super tune-up for us. Thanks everyone!

INDEX

A

A/C and P/S hoses installation, 157-158
Air conditioning compressor
  installation, 155
  removal, 20-21
Alternator
  bracket installation, 141
  installation, 156
  and power steering pump removal, 19-20
Assembly, see Engine assembly
Automatic transmission installation, 146-147

B

Balance test, power, 6-7
Battery removal, 17
BDC (bottom dead center), 40, 41, 61, 68
Bearing caps disassembly, 42-43
Belts installation, 156-157
Block
  casting numbers, 50
  inspection and reconditioning, 68-72
  preparation, 119
  serial number stampings, 49
Blowby, 2
Bore wear, 68
  dial-bore gauge or micrometer method, 70
  measuring, 68-71
  ring and feeler gauge method, 70
  sleeving, 69-70
  taper, 70-71
Bottom end disassembly, 39-45

C

Camshaft lobes
  wear, 11
  worn, 3
Camshafts
  disassembly, 43
  installation, 119-120
Camshafts and lifters
  cam bearing journals, 81
  checking the cams, 81
  design, 80-81
  inspection, 81
  run-out, 81
Carbon deposits, 3
Carburetor installation, 153
Casting numbers, 47-49, 50, 57
Chain stretch, 36
Chambered heads, 426 Hemispherical, 58
Cherry pickers, 14
Clutch
  and flywheel disassembly, 31
  mounting the, 149
  release bearing, 148-149
Clutch inspection, 147-149
  clutch release bearing, 148-149
  flywheel inspection, 148
  pressure plate inspection, 148
Coil installation, 156
Compression
  leaks, 66
  low, 3
  testing, 7
Compression rings installation, 125-127
Connecting rods, 54-55, 85-87
  interchangeability, 55
  parts and casting numbers chart, 55
  rod bolts, 87
Core plugs, 110
  disassembly, 44
  replacement, 93
Cranking vacuum test, 6
Crankshaft installation, 120-123
  laying the crankshaft, 121-123
  rear main seal, 121
Crankshaft interchange, 53-54
  383 cid, 53
  400 cid, 53
  413 and 426 cid Max Wedge, 53
  426 Hemi, 54
  440 cid, 53-54
Crankshaft pulley disassembly, 34-35
Crankshaft run-out, 78-79
  dial-indicator method, 78-79
  installation method, 78
Crankshaft torque converters, 53
Crankshafts, 52-54, 75-79
  checking after regrinding, 79
  cleaning, 79
  dampers, 53, 131
  inspecting main and connecting rod journals, 76-78
  laying the, 121-123
  matching rods, pistons and, 56-57
  pilot bushings, 79
  vibration damper chart, 54
CV (clearance volume), 55
Cylinder block specifications chart, 69
Cylinder bore finishing, 73-75
  chamfering the bores, 75
  cleaning the engine, 75
  main bearing caps, 74-75
  piston ring types, 73-74
Cylinder head assembly, 110
  core plugs, 110
  springs and valves, 110
Cylinder head casting numbers, B engine, 57
Cylinder head milling, 96-97, 97
Cylinder heads, 57-62, 132-135
Cylinder-wall wear versus operating temperatures, 68

D

Degreasers, 67
Detonation, 3
Diagnosis, 3-5
  checking spark plugs, 5-6
  internal noises, 3-5
  noises at engine speed, 4
  noises at half engine speed, 4-5
Dieseling, 3
Dipstick tube installation, 141
Disassembly, 29-45
  bottom end, 39-45
  camshaft, 43
  clutch and flywheel, 31
  core plugs, 44
  crankshaft pulley, 34-35
  distributor drive gear, 38
  draining the fluid, 30-31
  engine mounts, 32
  exhaust manifolds, 33
  flexplate, 31-32
  freeze plugs, 44
  front end, 33-36
  fuel pump, 33
  harmonic balancer, 34-35
  heads, 37-38
  intake manifolds, 37
  lifters, 38, 43-44
  main bearing caps, 42-43
  oil gallery plugs, 44
  oil pan, 39
  oil pump, 39
  procedures, 32-33
  removing exterior hardware, 30-32
  rocker arms, 37
  rod caps, 40-42
  rods and pistons, 40-42
  timing chain cover, 35
  timing chain and sprockets, 35-36
  top side, 36-38
  using engine stands, 30
  using ridge reamers, 38
  water pump housing, 33

windage tray, 40
Distributor
drive gear disassembly, 38
removal, 22
and spark plug wires installation, 154-155
Driveshafts and oil pumps, 87-89

E

Electrical components and wiring removal, 21-22
Electrical wiring installation, 158
alternator, 158
coil, 158
Engine
cleaning, 16
installation, 150-152
lifting, 150-151
Engine assembly, 115-143
camshaft installation, 119-120
crankshaft installation, 120-121
cylinder heads, 132-135
exterior hardware, 139
gaskets, 115-116
intake manifolds, 136-138
oil pan, 131-132
oil pump, 132
oil pump drives, 136
piston and connecting rod installation, 127-129
piston ring installation, 124-127
preparing the block, 119
sealers, 116
self-made gaskets, 117-118
timing chain cover, 129-131
timing chain and sprockets, 123-124
tools, 116-119
valve covers, 138
valvetrain, 135-136
Engine block identification, 50-52
413 cid Max Wedge, 51
426 cid Hemi, 52
426 cid Wedge, 51
440 cid, 51-52
B series 383 and 400 cid, 50-51
condensed history, 51
mounts, 51
RB series 413, 426 and 440 cid, 51-52
Engine block stamping numbers, 47
Engine break-in installation, 159-161
air cleaner, 161
cooling, 160
lubrication, 159-160
post break-in checks, 160-161
start-up, 160
Engine mounts
disassembly, 32
driver's side, 152
installation, 141
Engine removal, 13-27
air conditioning compressor, 20-21
alternator and power steering pump, 19-20
distributor, 22
electrical components and wiring, 21-22
exhaust pipes, 24-25
flexplate/torque converter, 24
getting started, 17-23
preparation, 15-17
radiator draining and removal, 17-19
radiator and water pump removal, 18-19
safety, 13-15
starter motor, 23-24
throttle linkage, vacuum lines & fuel lines, 22-23
tools, 15
top side parts, 23
under the car, 23-25
using the chain, 25-27
Engine stands, 30
EPA, 15, 30
Evans Speed Equipment, 72, 73
Exhaust manifolds
disassembly, 33
installation, 141
Exhaust pipes
installation, 152
removal, 24-25
Exhaust valves, burned, 3

F

Fan shroud installation, 158-159
Flexplate disassembly, 31-32
Flexplate/torque converter removal, 24
Fluid draining, 30-31
Flywheel
and clutch disassembly, 31
inspection, 148
Freeze plugs
disassembly, 44
replacement, 93
Fuel lines
and filters installation, 152
removal, 22-23
Fuel pumps
disassembly, 33
installation, 129

G

Gaskets, 115-116
and block deck surfaces, 66
removal, 66
self-made, 117-118
Grease and oil, 116

H

Hardware, exterior, 138-141
alternator bracket, 141
dipstick tube, 141
engine mounts, 141
exhaust manifolds, 141
spark plug wiring harnesses and heat shield, 141
thermostat, 138-141
throttle bellcrank, 141
throttle-spring bracket, 141
water pump housing, 138
water pump idler arm, 141
Harmonic balancer disassembly, 34-35
Head gaskets, 8-9
and block deck surfaces, 66
blown, 3
Head milling, 96-97
Head surface flatness, 96
Heads
chambered, 58
cylinder, 57-62
design fundamentals, 61-62
disassembly, 37-38
Head reconditioning, 91-113
cleaning and inspecting, 96-97
cylinder head assembly, 110
cylinder head milling, 96-97
disassembly, 92-97
head surface flatness, 96
intake manifolds, 111
rocker arm assemblies, 106-110
valve guides and stems, 97-101
valve seats, 100-101
valves disassembly, 92-93
Heater hose installation, 155
Hood removal, 16-17
Hot lash adjustment, 161
Hydraulic lifters, collapsed, 10-11

I

Identification numbers, 47-49
block casting numbers, 50
casting numbers, 47-49
engine block stamping numbers, 47

INDEX

Ignition, 62-63
Ignition timing, 162
Initial valve adjustment, 136
Inspection, 1-11
  checking the valvetrain, 9-11
  excessive oil consumption, 2
  piston rings, 2
  poor performance, 2-3
  valve guides, 2
Installation, 145-163
  A/C and P/S hoses, 157-158
  air conditioning compressor, 155
  alternator, 156
  automatic transmission, 146-147
  bellhousing to engine, 152
  belts, 156-157
  carburetor, 153
  coil, 156
  distributor and spark plug wires, 154-155
  driver side engine mount, 152
  electrical wiring, 158
  engine, 150-152
  engine break-in, 159-161
  exhaust pipes, 152
  fan shroud and radiator, 158-159
  flexplate to converter, 152
  fuel line and filter, 152
  heater hose, 155
  hot lash adjustment, 161
  lifting the engine, 150-151
  manual transmission, 147-149
  mounting the clutch, 149
  PCV valve, 153-154
  power steering pump, 155
  preparation, 149
  starter motor, 152
  top side, 152-159
  trial run, 161
  water pump and fan, 155
Intake manifolds, 97, 111, 136-138
  casting numbers chart, 59-60
  disassembly, 37

J

Journals
  checking finished, 78
  inspecting main and connecting rod, 76-78
  out of round, 76, 79
  surface finish, 76
  taper, 76-77
  tapered, 79
  thrust-face width, 77-78

L

Leak-down testing, 7-9
  head gasket, 8-9
  rings, 8
  valves, 9
Lifters, 81-82
  bores check, 73
  disassembly, 38, 43-44
  solid, 136
  worn, 3

M

Machining the block, 72
Magnaflux, 72, 73
Manifolds
  casting numbers, 59-60
  disassembly, 37
  exhaust, 141
  intake, 97, 111, 136-138
Manual transmission installation, 147-149
Max Wedge series, 57
Metal wear, 98
Mopar Performance, 51, 54, 63, 88, 96

N

Noises
  at engine speed, 4
  at half engine speed, 4-5
  internal, 3-5
Notching, 66

O

Oil consumption, excessive, 2
Oil gallery plugs disassembly, 44
Oil and grease, 116
Oil pans
  disassembly, 39
  installation, 131-132
Oil pressure, low, 2
Oil pumps
  disassembly, 39
  drives installation, 136
  installation, 132
Oil pumps and driveshafts, 87-89
  pressure relief valves, 89
  reassembling the oil pump, 89
Oil rings installation, 125

P

Parts identification and interchange, 47-63
PCV (positive crankcase ventilation), 2, 153-154
Performance, causes of poor, 2-3
  blown head gaskets, 3
  burned exhaust valves, 3
  carbon deposits, 3
  low compression, 3
  worn camshaft lobes, 3
  worn lifters, 3
Pilot bushings, 79
Piston ring installation, 124-127
  checking ring end gap, 124-125
  compression rings, 125-127
  oil rings, 125
Piston ring sets chart, 58
Piston rings, 2
Piston and rod assembly, 88
Piston-to-bore clearance, 71-72
  honing, 72
  knurling, 72
Pistons, 55-56, 82-85
  cast aluminum, 58
  checking and cleaning ring grooves, 83-85
  cranks and rods, matching, 56-57
  damage, 82-83
  disassembly and reassembly, 85
  pin bore wear, 83
  and rods disassembly, 40-42
  skirt diameter, 83
Plastigage, 122
Plugs disassembly, 44
Port and valve sizes, 61-62
Power steering pump
  installation, 155
  removal, 19-20
Preignition, 3
Pressure plate inspection, 148
Pressure relief valves, 89
Pushrods
  bent, 10
  installation, 135

R

Radiator
  draining and removal, 17-19
  installation, 158-159
  and water pump removal, 18-19
Regulations, EPA, 15
Ridge reamers, using, 38
Ring end gap, checking, 124-125
Ring grooves, 83-85
  compression, 84-85
  groove inserts, 85
  measuring width, 84
  oil, 85
Rings, 8

Rocker arm assemblies, 106-110, 135-136
426 Hemi, 108-110
adjustable rocker arms, 107-108
cleaning and inspection, 107-108
disassembly, 107
identifying left to right, 107
non-adjustable rockers, 108, 110
rocker arm shafts, 108
Rocker arms
disassembly, 37
failure, 10
shafts, 108
non-adjustable, 108, 110
Rod caps disassembly, 40-42
Rod interchangeability, connecting, 55
Rods
connecting, 54-55
cranks and pistons, matching, 56-57
and pistons assembly, 88
and pistons disassembly, 40-42

S

Safety, 13-15
Seal, rear main, 121
Sealers, 116
Serial number stampings, block, 49
Short block
bore wear, 68
camshafts and lifters, 79-82
cleaning, 66-67
cleaning and inspection, 65-68
connecting rods, 85-87
crankshaft, 75-79
crankshaft run-out, 78-79
cylinder bore finishing, 73-75
final block inspection and reconditioning, 68-72
gasket removal, 66
head gaskets and block deck surfaces, 66
machining the block, 72
oil pumps and driveshafts, 87-89
pistons, 82-85
reconditioning, 65-89
reconditioning, final checks, 72-73
thread chasing, 67-68
Spark plug wiring harnesses and heat shield installation, 141
Spark plugs, checking, 5-6
Spotcheck, 72, 73
Spring rate, 104
Springs and valves, 110
Starter motor
installation, 152
removal, 23-24
Stems and valve guides, 97-101
Studs, fixing stripped, 94-95
SV (swept volume), 61

T

Taper versus ring end-gap differences, 71
TDC (top dead center), 8, 55, 61, 135
Testing, 6-9
compression testing, 7
cranking vacuum test, 6
leak-down testing, 7-9
power balance test, 6-7
Thermostat installation, 138-141
Throttle
bellcrank installation, 141
linkage removal, 22-23
Throttle-spring bracket installation, 141
Timing chain cover, 129-131
crankshaft damper, 131
disassembly, 35
fuel pump, 129
oil seal, 129
Timing chain and sprockets, 123-124
cam and crank timing, 123
disassembly, 35-36
installing timing chain, 123-124
Timing, ignition, 162
Tools, 116-119
Top side disassembly, 36-38
Trial run, 161
Tune-up tips, 163

V

Vacuum lines removal, 22-23
Valve covers installation, 138
Valve floats, 103-104
Valve guide reconditioning, 99-100
knurling the guides, 98-99
oversized valves, 98
valve guide inserts, 98
Valve guides, 2
Valve guides and stems, 97-101
amount of guide clearance, 98-99
measuring valve guide wear, 97-99
valve guide reconditioning, 99-100
Valve inspection and reconditioning, 101-103
Valve jobs on high-mileage vehicles, 7
Valve lash chart, 136
Valve lifters installation, 135
Valve opening, checking
bent pushrods, 10
collapsed hydraulic lifters, 10-11
rocker arm failure, 10
Valve and port sizes, 61-62
Valve rocker shaft bracket locations, 109
Valve seats reconditioning, 100-101
grinding the seats, 100-101
specifications, 101
Valve spring inspection, 103-106
checking the springs, 105
free height, 104
installed spring height, 105-106
load at installed height, 104
load at open height, 104
solid height, 104
spring rate, 104
squareness, 105
valve float, 103-104
Valve springs
broken, 10-11
chart, 105
Valve stem
seals, 112-113
wear, 101-102
Valve terminology, 97
Valves, 9
disassembly, 92-93
hand-lapping, 100
oversized, 100
reconditioning, 102-103
sealing, 103
and springs, 110
Valvetrain, 103-104
broken valve spring, 10-11
camshaft lobe wear, 11
checking the, 9-11
initial valve adjustment, 136
installation, 135-136
pushrods, 135
rocker arm assemblies, 135- 136
solid lifters, 136
valve lifters, 135
Vehicle ID plate interpretation, 48

W

Warp check, 73
Water pump housings
chart, 62
disassembly, 33
installation, 138
Water pumps, 63
and fan installation, 155
idler arm installation, 141
removal, 18-19
Windage tray disassembly, 40

OTHER BOOKS FROM HPBOOKS AUTOMOTIVE

HANDBOOKS

Auto Electrical Handbook: 0-89586-238-7
Auto Upholstery & Interiors: 1-55788-265-7
Brake Handbook: 0-89586-232-8
Car Builder's Handbook: 1-55788-278-9
Street Rodder's Handbook: 0-89586-369-3
Turbo Hydra-matic 350 Handbook: 0-89586-051-1
Welder's Handbook: 1-55788-264-9

BODYWORK & PAINTING

Automotive Detailing: 1-55788-288-6
Automotive Paint Handbook: 1-55788-291-6
Fiberglass & Composite Materials: 1-55788-239-8
Metal Fabricator's Handbook: 0-89586-870-9
Paint & Body Handbook: 1-55788-082-4
Sheet Metal Handbook: 0-89586-757-5

INDUCTION

Holley 4150: 0-89586-047-3
Holley Carburetors, Manifolds & Fuel Injection: 1-55788-052-2
Rochester Carburetors: 0-89586-301-4
Turbochargers: 0-89586-135-6
Weber Carburetors: 0-89586-377-4

PERFORMANCE

Aerodynamics For Racing & Performance Cars: 1-55788-267-3
Baja Bugs & Buggies: 0-89586-186-0
Big-Block Chevy Performance: 1-55788-216-9
Big Block Mopar Performance: 1-55788-302-5
Bracket Racing: 1-55788-266-5
Brake Systems: 1-55788-281-9
Camaro Performance: 1-55788-057-3
Chassis Engineering: 1-55788-055-7
Chevrolet Power: 1-55788-087-5
Ford Windsor Small-Block Performance: 1-55788-323-8
Honda/Acura Performance: 1-55788-324-6
High Performance Hardware: 1-55788-304-1
How to Build Tri-Five Chevy Trucks ('55-'57): 1-55788-285-1
How to Hot Rod Big-Block Chevys:0-912656-04-2
How to Hot Rod Small-Block Chevys:0-912656-06-9
How to Hot Rod Small-Block Mopar Engines: 0-89586-479-7
How to Hot Rod VW Engines:0-912656-03-4
How to Make Your Car Handle:0-912656-46-8
John Lingenfelter: Modifying Small-Block Chevy: 1-55788-238-X
Mustang 5.0 Projects: 1-55788-275-4
Mustang Performance ('79–'93): 1-55788-193-6
Mustang Performance 2 ('79–'93): 1-55788-202-9
1001 High Performance Tech Tips: 1-55788-199-5
Performance Ignition Systems: 1-55788-306-8
Performance Wheels & Tires: 1-55788-286-X
Race Car Engineering & Mechanics: 1-55788-064-6
Small-Block Chevy Performance: 1-55788-253-3

ENGINE REBUILDING

Engine Builder's Handbook: 1-55788-245-2
Rebuild Air-Cooled VW Engines: 0-89586-225-5
Rebuild Big-Block Chevy Engines: 0-89586-175-5
Rebuild Big-Block Ford Engines: 0-89586-070-8
Rebuild Big-Block Mopar Engines: 1-55788-190-1
Rebuild Ford V-8 Engines: 0-89586-036-8
Rebuild Small-Block Chevy Engines: 1-55788-029-8
Rebuild Small-Block Ford Engines:0-912656-89-1
Rebuild Small-Block Mopar Engines: 0-89586-128-3

RESTORATION, MAINTENANCE, REPAIR

Camaro Owner's Handbook ('67–'81): 1-55788-301-7
Camaro Restoration Handbook ('67–'81): 0-89586-375-8
Classic Car Restorer's Handbook: 1-55788-194-4
Corvette Weekend Projects ('68–'82): 1-55788-218-5
Mustang Restoration Handbook('64 1/2–'70): 0-89586-402-9
Mustang Weekend Projects ('64–'67): 1-55788-230-4
Mustang Weekend Projects 2 ('68–'70): 1-55788-256-8
Tri-Five Chevy Owner's ('55–'57): 1-55788-285-1

GENERAL REFERENCE

Auto Math:1-55788-020-4
Fabulous Funny Cars: 1-55788-069-7
Guide to GM Muscle Cars: 1-55788-003-4
Stock Cars!: 1-55788-308-4

MARINE

Big-Block Chevy Marine Performance: 1-55788-297-5